DEFENSE, SECURITY AND STRATEGIES

TECHNOLOGY HORIZONS IN THE U.S. MILITARY

UNMANNED SYSTEMS AND AIR FORCE SCIENCE AND TECHNICAL ENDEAVORS

DEFENSE, SECURITY AND STRATEGIES

Additional books in this series can be found on Nova's website under the Series tab.

Additional E-books in this series can be found on Nova's website under the E-book tab.

DEFENSE, SECURITY AND STRATEGIES

TECHNOLOGY HORIZONS IN THE U.S. MILITARY

UNMANNED SYSTEMS AND AIR FORCE SCIENCE AND TECHNICAL ENDEAVORS

ETHAN J. JOHNSON
AND
GABRIELLE L. RODRIGUEZ
EDITORS

Nova Science Publishers, Inc.
New York

For permission to use material from this book please contact us:
Telephone 631-231-7269; Fax 631-231-8175
Web Site: http://www.novapublishers.com

NOTICE TO THE READER

The Publisher has taken reasonable care in the preparation of this book, but makes no expressed or implied warranty of any kind and assumes no responsibility for any errors or omissions. No liability is assumed for incidental or consequential damages in connection with or arising out of information contained in this book. The Publisher shall not be liable for any special, consequential, or exemplary damages resulting, in whole or in part, from the readers' use of, or reliance upon, this material. Any parts of this book based on government reports are so indicated and copyright is claimed for those parts to the extent applicable to compilations of such works.

Independent verification should be sought for any data, advice or recommendations contained in this book. In addition, no responsibility is assumed by the publisher for any injury and/or damage to persons or property arising from any methods, products, instructions, ideas or otherwise contained in this publication.

This publication is designed to provide accurate and authoritative information with regard to the subject matter covered herein. It is sold with the clear understanding that the Publisher is not engaged in rendering legal or any other professional services. If legal or any other expert assistance is required, the services of a competent person should be sought. FROM A DECLARATION OF PARTICIPANTS JOINTLY ADOPTED BY A COMMITTEE OF THE AMERICAN BAR ASSOCIATION AND A COMMITTEE OF PUBLISHERS.

Additional color graphics may be available in the e-book version of this book.

Library of Congress Cataloging-in-Publication Data

ISBN: 978-1-62081-364-5

Published by Nova Science Publishers, Inc. † New York

CONTENTS

PREFACE

Air Force warfighting capabilities have a proud heritage of being born from the very best science and technology our Nation can create; indeed, the very history of the United States Air Force is closely intertwined with the development of advances in science and technology. Yet today, "flattening" of the world is making it increasingly challenging for the U.S. to maintain technology superiority over potential adversaries. A growing number of nations soon will have the ability to transform science and technology advances into militarily significant capabilities. Over the next decades, we will increasingly face potential adversaries having peer or near-peer capabilities. This book examines the key areas of science and technology that the Air Force must focus on over the next two decades to enable technologically achievable capabilities that can provide it with the greatest U.S. Joint force effectiveness by 2030.

Chapter 1 - U.S. and allied combat operations continue to highlight the value of unmanned systems in the modern combat environment. Combatant Commanders (CCDRs) and warfighters value the inherent features of unmanned systems, especially their persistence, versatility, and reduced risk to human life. The U.S. military Services are fielding these systems in rapidly increasing numbers across all domains: air, ground, and maritime. Unmanned systems provide diverse capabilities to the joint commander to conduct operations across the range of military operations: environmental sensing and battlespace awareness; chemical, biological, radiological, and nuclear (CBRN) detection; counter-improvised explosive device (C-IED) capabilities; port security; precision targeting; and precision strike. Furthermore, the capabilities provided by these unmanned systems continue to expand.

The Department of Defense (DoD) has been successful in rapidly developing and fielding unmanned systems. DoD will continue to focus on responding rapidly to CCDR requirements, while ensuring systems are acquired within the framework of DoD's new wide-ranging Efficiencies Initiatives[1]. In the fiscal environment facing the Nation, DoD, in concert with industry, must pursue investments and business practices that drive down life-cycle costs for unmanned systems. Affordability will be treated as a key performance parameter (KPP) equal to, if not more important than, schedule and technical performance. DoD will partner with industry to continue to invest in unmanned systems technologies while providing incentives for industry to implement cost-saving measures and rewarding industry members that routinely demonstrate exemplary performance.

Chapter 2 - From its inception, the Air Force has undertaken a major effort roughly once every decade to articulate a vision for the science and technology (S&T) advancements that it should undertake to achieve over the following decade to enable the capabilities that it will need to prevail. Six such S&T visions have been developed, beginning with "Toward New Horizons" in 1945 led by Theodore von Kármán for Gen Hap Arnold, through "New World Vistas" conducted in 1995.

Since completion of the latter, 15 years have passed without an updated Air Force S&T vision. "Technology Horizons" represents the next in this succession of major vision efforts conducted at the Headquarters Air Force level. In view of the far-reaching strategic changes, rapid global technological advances, and growing resource constraints over the next decade, this is an overdue effort that can help guide S&T investments to maximize their impact for maintaining Air Force technological superiority over potential adversaries.

In: Technology Horizons in the U.S. Military
Editors: Ethan J. Johnson and Gabrielle L. Rodriguez
ISBN 978-1-62081-364-5

Chapter 1

UNMANNED SYSTEMS INTEGRATED ROADMAP FY2011-2036*

Department of Defense

EXECUTIVE SUMMARY

U.S. and allied combat operations continue to highlight the value of unmanned systems in the modern combat environment. Combatant Commanders (CCDRs) and warfighters value the inherent features of unmanned systems, especially their persistence, versatility, and reduced risk to human life. The U.S. military Services are fielding these systems in rapidly increasing numbers across all domains: air, ground, and maritime. Unmanned systems provide diverse capabilities to the joint commander to conduct operations across the range of military operations: environmental sensing and battlespace awareness; chemical, biological, radiological, and nuclear (CBRN) detection; counter-improvised explosive device (C-IED) capabilities; port security; precision targeting; and precision strike. Furthermore, the capabilities provided by these unmanned systems continue to expand.

The Department of Defense (DoD) has been successful in rapidly developing and fielding unmanned systems. DoD will continue to focus on responding rapidly to CCDR requirements, while ensuring systems are acquired within the framework of DoD's new wide-ranging Efficiencies Initiatives[1]. In the fiscal environment facing the Nation, DoD, in concert with industry, must pursue investments and business practices that drive down life-cycle costs for unmanned systems. Affordability will be treated as a key performance parameter (KPP) equal to, if not more important than, schedule and technical performance. DoD will partner with industry to continue to invest in unmanned systems technologies while providing incentives for industry to implement cost-saving measures and rewarding industry members that routinely demonstrate exemplary performance.

> My Acquisition Decision Memorandum (ADM) approving formal program commencement of the program will contain an affordability target to be treated by

* This is an edited, reformatted and augmented version of the Department of Defense, 11-S-3613.

the Program Manager like a Key Performance Parameter (KPP) such as speed, power, or data rate … .

–Under Secretary of Defense Memorandum for Acquisition Professionals, Better Buying Power, September 2010[1]

This document provides a DoD vision for the continuing development, fielding, and employment of unmanned systems technologies. Since publication of the last DoD Roadmap in 2009, the military Services have released individual Service roadmaps or related strategy documents. This roadmap defines a common vision, establishes the current state of unmanned systems in today's force, and outlines a strategy for the common challenges that must be addressed to achieve the shared vision.

The challenges facing all military Services in the Department include:

1) Interoperability: To achieve the full potential of unmanned systems, these systems must operate seamlessly across the domains of air, ground, and maritime and also operate seamlessly with manned systems. Robust implementation of interoperability tenets will contribute to this goal while also offering the potential for significant life-cycle cost savings.
2) Autonomy: Today's iteration of unmanned systems involves a high degree of human interaction. DoD must continue to pursue technologies and policies that introduce a higher degree of autonomy to reduce the manpower burden and reliance on full-time high-speed communications links while also reducing decision loop cycle time. The introduction of increased unmanned system autonomy must be mindful of affordability, operational utilities, technological developments, policy, public opinion, and their associated constraints.
3) Airspace Integration (AI): DoD must continue to work with the Federal Aviation Administration (FAA) to ensure unmanned aircraft systems (UAS) have routine access to the appropriate airspace needed within the National Airspace System (NAS) to meet training and operations requirements. Similar efforts must be leveraged for usage of international airspace.
4) Communications: Unmanned systems rely on communications for command and control (C2) and dissemination of information. DoD must continue to address frequency and bandwidth availability, link security, link ranges, and network infrastructure to ensure availability for operational/mission support of unmanned systems. Planning and budgeting for UAS Operations must take into account realistic assessments of projected SATCOM bandwidth, and the community must move toward onboard pre-processing to pass only critical information.
5) Training: An overall DoD strategy is needed to ensure continuation and Joint training requirements are in place against which training capabilities can be assessed. Such a strategy will improve basing decisions, training standardization, and has the potential to promote common courses resulting in improved training effectiveness and efficiency.
6) Propulsion and Power: The rapid development and deployment of unmanned systems has resulted in a corresponding increased demand for more efficient and logistically supportable sources for propulsion and power. In addition to improving system

effectiveness, these improvements have the potential to significantly reduce life-cycle costs.

7) Manned-Unmanned (MUM) Teaming: Today's force includes a diverse mix of manned and unmanned systems. To achieve the full potential of unmanned systems, DoD must continue to implement technologies and evolve tactics, techniques and procedures (TTP) that improve the teaming of unmanned systems with the manned force.

This Roadmap leverages individual Service roadmaps and visions, and identifies challenges that might stand in the way of maturing those visions to a shared Joint vision. The vignettes provided at the beginning of the Roadmap give the reader a glimpse into potential unmanned systems capabilities. They do not serve as requirements—the individual Services will continue to identify requirements gaps and utilize the Joint Capabilities Integration and Development System (JCIDS) to determine which requirements to fund. The chapters that follow the vignettes identify core areas that are challenges for further growth in unmanned systems and chart out science, technology, and policy paths that will enable unmanned systems to fulfill an expanding role in supporting the warfighter. Success in each of these areas is critical to achieve DoD's shared vision and realize the full potential of unmanned systems at an affordable cost.

> ... the ability to understand and control future costs from a program's inception is critical to achieving affordability requirements.
>
> *–Under Secretary of Defense Memorandum for Acquisition Professionals, Better Buying Power, September 2010*[1]

1. Introduction/Scope

1.1. Purpose

The purpose of this document is to describe a vision for the continued integration of unmanned systems into the Department of Defense (DoD) Joint force structure and to identify steps that need to be taken to affordably execute this integration. DoD has seen rapid growth, sparked in large part by the demands of the current combat environment, in the development, procurement, and employment of unmanned systems. Today's deployed forces have seen how effective unmanned systems can be in combat operations. This experience has created expectations for expanding the roles for unmanned systems in future combat scenarios. This Roadmap establishes a vision for the next 25 years and outlines major areas where DoD and industry should focus to ensure the timely and successful adoption of unmanned systems.

1.2. Scope

This Roadmap follows the path originally laid out in the 2007 and 2009 Roadmaps in addressing all three unmanned domains: air, ground, and maritime. However, this document deviates from the earlier editions, primarily as a result of the following:

- An Unmanned Systems Roadmap survey conducted by the Office of the Under Secretary of Defense (Acquisition, Technology, and Logistics) (OUSD(AT&L))
- Publication of service-specific roadmaps for unmanned systems

Shortly after the publication of the 2009 Roadmap, OUSD(AT&L) conducted a survey of key stakeholders and users of the Roadmap. The survey sampled a wide audience, including: Office of the Secretary of Defense (OSD), Service headquarters, warfighting commands, Service acquisition organizations, Service laboratories, multiple Joint organizations, other government agencies, industry (both large and small businesses), and academia. One of the major outcomes of this survey was a decision to capture the catalog function of the Roadmap in a separate, online tool. The reason for this decision is that the online tool provides greater functionality than that of the two-dimensional, hard-copy catalog, including the capability for more frequent updates than the biennial printed Roadmap. The catalog can be found on the Unmanned Warfare Information Repository site at: https://extranet. acq.osd.mil/uwir/.

The survey also helped define the audience for the 2011 edition. This Roadmap provides a common vision and problem set to help shape military Service investments. The document also describes DoD's direction to help industry participants shape their investments, particularly with respect to independent research and development.

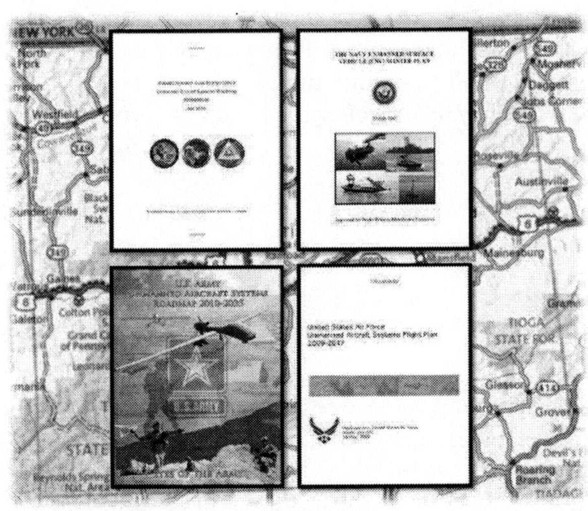

Since the publication of the 2009 Roadmap, each military Service has developed its own roadmap or equivalent document (listed in Appendix A). The U.S. Air Force (USAF) released its "Unmanned Aircraft Systems Flight Plan" in 2009 outlining an actionable plan across the diverse spectrum of doctrine, organization, training, materiel, leadership and education, personnel, facilities, and policy (DOTMLPF-P). In 2009, the U.S. Army published the "Unmanned Ground Systems Roadmap," providing a common resource document for Army and U.S. Marine Corps (USMC) stakeholders in unmanned ground vehicles (UGVs). The Army released its unmanned aircraft systems (UAS) Roadmap in 2010, which established a broad vision for developing, organizing, and employing UAS across the spectrum of Army operations. In November 2009, the USMC published its "Concept of Operations for USMC Unmanned Aircraft Systems Family of Systems (CONOPS for USMC UAS FoS)." Finally, the U.S. Navy published its "Information Dominance Roadmap for Unmanned Systems" in December 2010. In light of these recent publications (see right), this Roadmap was tailored to focus on common issues facing all Services as well as to articulate a vision for achieving these goals in today's fiscal environment. The goal for this document is to serve as a single, unified source to clearly articulate the DoD common vision for unmanned systems and to identify a common problem set facing DoD in maximizing the military utility offered by these versatile and innovative systems.

2. VISION

The Department of Defense's vision for unmanned systems is the seamless integration of diverse unmanned capabilities that provide flexible options for Joint Warfighters while exploiting the inherent advantages of unmanned technologies, including persistence, size, speed, maneuverability, and reduced risk to human life. DOD envisions unmanned systems seamlessly operating with manned systems while gradually reducing the degree of human control and decision making required for the unmanned portion of the force structure.

> ... to ensure safe, effective and supportable capabilities are provided while meeting cost, schedule and performance. The parallel vision is to provide continuous improvement of unmanned system capabilities to meet current and future Warfighter objectives.
>
> *– Mission and Vision, Robotic Systems Joint Project Office Unmanned Ground Systems Roadmap, July 2009*

> ... develop and field cost-effective USVs to enhance Naval and Joint capability to support: Homeland Defense, the Global War on Terror, Irregular Warfare, and conventional campaigns.
>
> *– The USV vision, The Navy Unmanned Surface Vehicle Master Plan, 23 July 2007*

> ... adopt innovative strategies to provide cost effective logistical support ...
>
> *– Goals and Objectives, US Army Roadmap for UAS 2010-2035*

> ... to harness increasingly automated, modular, globally connected, and sustainable multi-mission unmanned systems resulting in a leaner, more adaptable and efficient Air Force that maximizes our contribution to the Joint Force.
>
> *– USAF vision, USAF Unmanned Aircraft Systems Flight Plan 2009-2047*

2.1. Future Operational Environment

The strategic environment and the resulting national security challenges facing the United States for the next 25 years are diverse. The United States faces a complex and uncertain security landscape in which the pace of change continues to accelerate. The rise of new powers, the growing influence of nonstate actors, the spread of weapons of mass destruction and other irregular threats, and continuing socioeconomic unrest will continue to pose profound challenges to international order.

Over the next two decades, forces will operate in a geostrategic environment of considerable uncertainty with traditional categories of conflict becoming increasingly blurred. This era will be characterized by protracted confrontation among state, nonstate, and individual actors using violent and nonviolent means to achieve their political and ideological goals. Future adversaries will rely less on conventional force-on-force conflicts to thwart U.S. actions and more on tactics that allow them to frustrate U.S. intentions without direct confrontation.

> The future operating environment will be one of constant and accelerating change. Economic, demographic, resource, climate, and other trends will engender competition locally, regionally, and globally.... State and non-state actors will find new and more deadly means of conducting operations in all domains, to include land, air, maritime, and cyberspace to further their aims ... otherwise leveraging land, air, and maritime areas to ensure their freedom of movement and deny it to others..
>
> – *Joint Operational Concept, Irregular Warfare: Countering Irregular Threats*

As technological innovation and global information flows accelerate, nonstate actors will continue to gain influence and capabilities that, during the past century, remained largely the purview of states. Chemical and biological agents will become increasingly more accessible, lethal and sophisticated. Both state and nonstate actors will actively pursue nuclear weapons, sophisticated and/or bioengineered biological agents, and nontraditional chemical agents.

> The next quarter century will challenge U.S. Joint Forces with threats and opportunities ranging from regular and irregular wars in remote lands, to relief and reconstruction in crisis zones, to cooperative engagement in the global commons .... There will continue to be opponents who will try to disrupt the political stability and deny free access to the global commons that is crucial to the world's economy.... In this environment, the presence, reach, and capability of U.S. military forces, working with like-minded partners, will continue to be called upon to protect our national interests.
>
> – *Joint Operating Environment 2010: Ready For Today, Preparing For Tomorrow*

Unmanned systems can help in countering these threats by reducing risk to human life and increasing standoff from hazardous areas.

2.2. DoD's Vision

The DoD, along with industry, understands the effect that innovation and technology in unmanned systems can have on the future of warfare and the ability of the United States to adapt to an ever-changing global environment. DoD and industry are working to advance operational concepts with unmanned systems to achieve the capabilities and desired effects on missions and operations worldwide. In building a common vision, DoD's goals for unmanned systems are to enhance mission effectiveness, improve operational speed and efficiency, and affordably close warfighting gaps.

DoD is committed to harnessing the potential of unmanned systems and strengthening mission effectiveness while maintaining fiscal responsibility. DoD will also work on establishing a complementary relationship between manned and unmanned capabilities while optimizing commonality and interoperability across space, air, ground, and maritime domains.

Open architecture (OA) and open interfaces need to be leveraged to address problems with proprietary robotic system architectures. Standards and interface specifications need to be established to achieve modularity, commonality, and interchangeability across payloads, control systems, video/audio interfaces, data, and communication links. This openness will enhance competition, lower life-cycle costs, and provide warfighters with enhanced unmanned capabilities that enable commonality and joint interoperability on the battlefield.

By prudently developing, procuring, integrating, and fielding unmanned systems, DoD and industry will ensure skillful use of limited resources and access to emerging warfighting capabilities. Pursuing this approach with unmanned systems will help DOD sustain its dominant global military power and provide the tools required by national decision-makers to influence foreign and domestic activities while adapting to an ever-changing global environment. The following quotation captures the breadth of the challenge:

> I speak for the Navy, that unmanned systems have to address all of the domains in which the Navy operates.... We operate on the surface, above the surface, into space, but then we operate below the surface. So when we talk about unmanned and ... as we knit all of this capability together and capacity together, it has to take into account that we're operating in all those different domains.
>
> – *Admiral Gary Roughead, Chief of Naval Operations (CNO)*

With the current fiscal environment of constrained budgets, affordability is a factor across the entire acquisition cycle and must be actively engaged by the program managers, users, trainers, and testers to identify problems early, and address cost throughout the life cycle. A dollar saved early results in hundreds of saved dollars compared to problems resolved in production or worse yet during operations and support. While "open systems architecture and data rights" are critical to keeping costs in check, emphasizing the removal of obstacles to competition and the opportunities in test and evaluation (T&E) to facilitate competitive analysis is equally important to reduce developmental costs.

The assembly line of activity involved in producing unmanned systems must address risk across the life cycle to address the new challenges of testing autonomous functionality in the initial stages, and evaluating the operation and support issues involved in sustainment for increasing reliability, availability, and maintainability. The emphasis on vignettes at a mission level only indirectly emphasizes the increasing need for an evolutionary capability in unmanned systems production that is resilient and responsive to the dynamic situation faced by today's warfighter. New technology, methodologies, and human resourcing are critical for establishing rapid acquisition environments that maximize the potential for unmanned systems production.

2.3. Vignettes

The following vignettes offer examples of the increased capability and flexibility inherent in unmanned systems as DoD continues to field unmanned technologies and integrate resulting systems into its existing force structure. These vignettes are not intended to present an exhaustive list of the possibilities, but rather to present a few examples to illustrate the vision described throughout this Roadmap.

2.3.1. Interoperability Across Domains Vignette, 2030s

Location: Northern Pacific Littoral Areas

Situation: The number and boldness of coordinated, provocative efforts between the Republic of Orangelandia (ROO) and the increasing number of radicalized Islamic nation-states within the tropic zones (± 20° latitude) have increased over the past 15 years. ROO has demonstrated a delivery capability for nuclear intercontinental ballistic missiles, and several radical Islamic nations now openly possess nuclear weapon technology. Although nuclear power's role is expanding, oil remains the energy resource of preference even though gaining access to oil by Western nations has become increasingly constrained and expensive. The United States' gross domestic product (GDP) is being challenged by China.

Scenario: A 50-year-old, former Soviet-era, *Akula* class, nuclear-powered attack submarine sails out of ROO's Molan harbor at night unobserved by Western reconnaissance satellites. Movements of ROO submarines are of high interest due to their rarity (fewer than a dozen occurrences a year) and primarily due to ROO's status as a rogue, nuclear-capable nation-state. The submarine's departure is detected by the underwater surveillance grid, which is monitoring vessel movements in and out of the ROO waters. Ahead of the submarine, a glider unmanned underwater vehicle (UUV) is autonomously detached from the local network to intercept the faster submarine. Closing to within 50 yards as the submarine passes, the UUV succeeds in attaching a tether to the submarine, which begins pulling the UUV along (see figure right). As the submarine dives below the UUV's operating depth, the UUV adjusts the tether to maintain its position close to the surface. Every three hours, it glides to the surface and transmits a low-power position report.

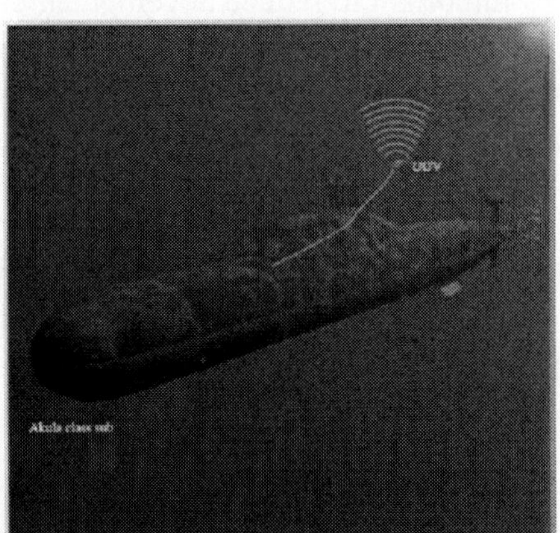

The position reports are received by an orbiting communications relay, Baton One, an EQ-25 UAS operating at 75,000 ft in the eastern Pacific region. The EQ-25 is an extreme-endurance UAS, capable of operating for two months on station without refueling. As the submarine enters the Sea of Okhotsk and heads toward the North Pacific Ocean, U.S. military commanders are faced with a decision. Despite the advanced battery technology of the UUV, the battery life is finite; therefore, the operators have three courses of action affecting their surveillance operation: (1) continue surveillance by shifting the orbit of Baton One to maintain reception range on the UUV, which will otherwise be lost in 12 hours (2) save the UUV by detaching it when its remaining power is still sufficient for it to recover itself (within three days) or (3) expend the UUV by keeping it attached until its power is exhausted (within six days). Because ROO submarines seldom sortie beyond the littoral seas of northeastern Asia, they decide to shift Baton One's orbit and wait to decide the UUV's fate until the submarine's intent becomes clearer.

By the third day, the submarine is heading toward the mid-Pacific and the Hawaiian Islands. Because the value of the mission exceeds the cost of the asset, the decision is made to expend the low-cost UUV to buy time for a naval anti-submarine warfare (ASW) ship to intercept and track the submarine. The following day, the submarine reverses course. Two days later, the still-attached UUV converts to beacon mode to conserve its dwindling power reserves, Baton One returns to its planned orbit, and the ASW ship turns for Pearl Harbor. A Broad Area Maritime Surveillance (BAMS) UAS, MQ-4C, is launched from Guam to track the beacon. It recovers the beacon's signal and determines that it is stationary. Autonomously descending with its internal airborne sense and avoid (ABSAA) system to maintain "due regard," the BAMS UAS is able to visually acquire the UUV, floating in mid-ocean and no longer attached to the ROO submarine—potentially detached by the submarine's crew. The submarine's position and intent are now unknown.

Ten days later, a weak seismic disturbance is detected 150 miles southeast of Anchorage, Alaska (see map below). Several minutes later, a much more significant event registers 3.5 on the Richter scale. An interagency DoD/homeland defense reconnaissance UAS is launched out of Elmendorf Air Force Base (AFB) and detects a radiation plume emanating near Montague Island at the mouth of Prince William Sound. The UAS maps the plume as it begins spreading over the sound, and a U.S. Coast Guard offshore patrol cutter deployed from Kodiak employs its embarked unmanned helicopter to drop buoys with chemical, biological, radiological, and nuclear (CBRN) sensors in the Sound and within narrow passes to measure fallout levels. The plume begins to spread over the sound and threatens the city of Valdez. All vessel traffic, mainly oil tankers, transiting in and out of the Sound is stopped, and operations at the oil terminal are suspended. Oil storage facilities at the terminal are quickly filled to capacity, and the flow from Prudhoe Bay is shut down. The port of Valdez, the largest indigenous source of oil for the United States, is effectively under quarantine.

Due to the growing contamination of the local environment, disaster response officials decide to request the support of the military because of their experience both with operations in CBRN zones and with unmanned systems, which are the tools of choice because of the contamination hazards to personnel. The amphibious transport dock ship USS *New York* anchors near an entrance to Prince William Sound and begins operations with its unmanned surface vehicles (USVs) and MQ-8 detachments. An EQ-25 orbit is established over the Sound to ensure long-term, high-volume communication capability in the high-latitude, mountainous region. With data compression technology fielded in the transmitting and relay systems, the EQ-25 is capable of handling all the theater data relay requirements. A USV proceeds to the focus of contamination and lowers a tethered remotely operated vehicle (ROV) to conduct an underwater search for the source. The USV's sonar quickly locates a large object in very shallow water and, on closer inspection by the ROV, images the severely damaged hull of what appears to be an *Akula* class submarine. The hull is open to the sea, and the ROV places temperature gradient sensors on the hull and inserts gamma sensors into the exposed submarine compartments. The Joint Task Force that was formed to manage the disaster quickly determines that the reactor fuel core is exposed to the sea and that the reactor was not shut down and is still critical. Suspicion of the submarine's origin centers on its being from ROO, but all evidence that this vessel is the lost *Akula* submarine is currently circumstantial.

The radiation plume has now encompassed the evacuated town of Valdez, and MQ-8s fly repeated sorties to the town, dock, and terminal areas to deploy UGVs with sensors and collect samples for analysis. Returning USVs and MQ-8s are met and serviced by personnel in hazmat gear and washed down after each sortie. With conditions deteriorating, two unmanned Homeland Defense CBRN barges fitted with cranes, containers, and remote controls arrive from Seattle. USVs are stationed in the narrow straits leading into the Sound with hydrophones to broadcast killer whale sounds to frighten fish outside the Sound away from the contaminated area. Over the next two weeks, with the assistance of U.S and coalition ROVs equipped with cutting torches, grappling fixtures, and operating from USVs, one remotely operated submersible barge is able to work around the clock with impunity against exposure levels to recover the exposed fuel sources and to isolate them in specially designed containers. A second barge similarly retrieves sections of the crippled submarine. Both barges operate with a high degree of autonomy, limiting exposure of personnel to the radioactive contamination.

The UGVs continue monitoring contamination levels and collecting samples, but now also start conducting decontamination of the oil terminal control station and the local power and water facilities. Highly contaminated soil is placed into steel drums, and larger UGVs are used to dig pits and bury contaminated building and pipeline materials. Advanced sensor technology and control logic allows the UGVs to operate around the clock with human operators serving solely in a monitoring function. USVs are used to collect carcasses floating in the Sound and bring them to shore for disposal. UUVs crisscross the seafloor of the Sound to locate and tag remnants of the submarine for later collection. Unmanned aircraft (UA) fly continuously through the National Airspace System (NAS) at low altitude to monitor and map the declining radiation contours, at medium altitude to map cleanup operations, and at high altitude to relay control commands and data from the nearly one hundred unmanned vehicles at work. Decontamination, refueling, and repair shops have been established in nearby Cordova to service the vehicles and aircraft and on the USS *New York* to service the boats and submersibles. It is the largest coordinated use of international air, ground, and maritime unmanned systems ever conducted.

2.3.2. African Maritime Coalition Vignette, 2030s

Location: Gulf of Guinea off the coast of Africa

Situation: An UAS and an UUV, deployed from littoral combat ship (LCS) *Freedom*, are on patrol monitoring the littoral oil infrastructure of a developing nation-state. This nation-state has recently adjusted its geopolitical stance to ally itself militarily and economically with the United States and friendly European governments.

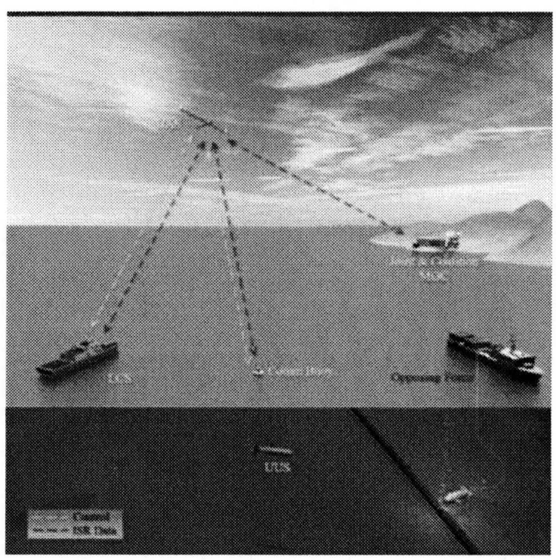

Scenario: The *Freedom*'s UUV in its assigned patrol area detects an anomaly, a remote pipeline welder controlled by an unknown force. The underwater remote welder is positioning itself to intersect a major underwater oil pipeline. Using its organic "smart software" processing capability, the UUV evaluates the anomaly as a possible threat and releases a communications buoy that transmits an alert signal and a compressed data "snapshot" from the UUV's onboard video/acoustic sensor.

The communications buoy's low probability of intercept (LPI) data are relayed via a small tactical unmanned aircraft system (STUAS) to other units in the area and to the Joint Maritime Operations Center (JMOC) ashore. The commander on the LCS directs the UUV and the UAS to provide persistent intelligence, surveillance, and reconnaissance (ISR) and command and control (C2) relay support. Simultaneously, the UAS transmits corroborating ISR data on a suspect vessel near the UUV anomaly. Thanks to a recently fielded, advanced technology propulsion upgrade, the STUAS is able to stay on station for 24 hours before being relieved (see graphic right).

Meanwhile, the JMOC analysts recognize the pipeline welder in the UUV data snapshot as one recently stolen and acquired by rebel antigovernment forces. The JMOC then dispatches an Allied quick reaction force (QRF) via 160th Special Operations Aviation Regiment (SOAR) aircraft and USAF CV-22 Osprey from a nearby airfield. The JMOC retasks a special warfare combatant-craft crewman (SWCC) Mk V to investigate and neutralize the potential hostile surface vessel controlling the stolen pipeline welder. The SWCC Mk V launches its own small UA to provide a low-level ISR view ahead of its navigation track while providing an LPI secure communications path among the special forces QRF team. The SWCC Mk V's UA provides a real-time common data link (CDL) common operational picture (COP) data stream via the higher altitude UAS to the LCS and JMOC.

The JMOC receives a signals intelligence (SIGINT) alert that the suspect hostile surface vessel is launching a Russian Tipchak, a medium-altitude, long-endurance (MALE) UA. The latest Tipchak variant is a hybrid UA with US-derived systems and avionics. This Tipchak is capable of launching short-range air-to-air missiles (AAMs) or air-to-surface missiles (ASMs). Its host platform, the suspect hostile vessel, has an early warning suite and has probably detected the LCS nearby or visually sighted the SWCC Mk V's UA. An update to the SIGINT alert at the JMOC reveals the Tipchak is being launched for a surveillance sweep and counter-air/counter-UA mission.

Realizing the hostile UA could pose a risk or even jeopardize the QRF, the JMOC commander launches a USAF MQ-1000 UA optimized for air interdiction and ground strike. The MQ-1000 UA, empowered by rules of engagement (ROE) allowing autonomous operation, immediately conducts an air-to-air engagement and neutralizes the Tipchak UA.

The SWCC Mk V's special forces team then conducts a visit, board, search, and seizure (VBSS) on the suspected hostile vessel supporting the UUV pipeline interdictor. Since the threat is neutralized, the unmanned systems update their patrol status, cancel the alert status, and recover or resume their assigned patrol sectors.

2.3.3. Complex Unmanned Systems Test and Evaluation Scenario

As unmanned systems become more complicated, more integrated, more collaborative, and more autonomous, establishing test-driven development constructs and infrastructure for supporting early-onset test and evaluation (T&E) and life-cycle T&E will become increasingly critical. The Weapon Systems Acquisition Reform Act (WSARA) of 2009 sets the stage for advancing T&E that will address cost saving through early-on engagement and effective sustainment in facilitating unmanned systems acquisition.

> "The key to successful acquisition programs is getting things rightfromthe startwith sound systems engineering, cost estimating, and developmental testing early in the program cycle. The bill that we are introducing today will require the Department of Defense to take the steps needed to put major defense acquisition programs on a Sound footing from the outset. If these changes are successfully implemented, they should help our acquisition programs avoid future cost overruns, schedule delays, and performance problems."
>
> *–Senator Carl Levin, Chairman, Senate Armed Services Committee*

> "The Weapon System Acquisition Reform Act of 2009 is an important step in efforts to reform the defense acquisition process. This legislation is needed to focus

acquisition and procurement on emphasizing systems engineering; more effective upfront planning and management of technology risk; and growing the acquisition workforce to meet program objectives."

–Senator John McCain, Ranking Member,
Senate Armed Services Committee

The two previous vignettes have focused on the utility of unmanned systems, but it is also helpful to focus on unmanned systems acquisition. The acquisition of systems with increasing net-centricity and automated functionality will introduce unexpected levels of risk. Systems engineering involves decomposing a design into separable elements, characterizing the intended relationships between them, and verifying the system built to specification operates as intended. The systems engineering "V" represents decomposition and design on the downstroke and integration, verification, and validation on the upstroke. As systems become more complex, the difficulties of addressing the upstroke of the "V" increase. T&E is critical for addressing this risk. The systems engineering of complex systems is gets scrutinized when major problems lead to program delays, cost overruns, and even cancellations. The issues typically lie with unintended and unanticipated interactions between elements that are uncovered only during integration, testing, or once in service. The T&E of manned systems has created optimal strategies for reducing risk in the areas of frequency, performance, support systems, and safety. The primary challenge for today's defense acquisition system is to execute acquisition programs without major schedule delays and cost overruns. Meeting that challenge has been the goal of acquisition reform improvements for decades.

Unmanned systems raise new issues of artificial intelligence, communications, autonomy, interoperability, propulsion and power, and manned-unmanned (MUM) teaming that will challenge current T&E capabilities. These problems will get more serious as systems become more interactive and more automated. Failures often occur at the interfaces between system elements, in many cases, between interfaces thought to be separate. The exponential trends in software and network communications increasingly mean that many elements of a system can now affect one another. The incredible complexity of millions of lines of software requires new approaches for detecting problems earlier in the design phase where cost mitigation is most effective. As systems get much of their functionality from software and multisystem interactions, complexity is no longer separate and distinct. Complexity is about the whole ecosystem, and systems engineering has to become more holistic. Model-based systems engineering, already in use by the software and circuit industries, is augmenting document-driven approaches in important ways. Executable models can be effective conveyors of information throughout a supply chain. Models designed to provide contextual information about the degrees of freedom and the interactions could potentially pass from the Government to the prime supplier and on to second-and third-tier suppliers from early concept through Milestones A, B, and C into the operations and support phase.

Location: DoD T&E Centers Across the World

Situation: In a rapid acquisition support environment, integrated T&E teams work with trainers and the user to accelerate the production of unmanned systems. The model involves industries that must leverage test frameworks to take advantage of Moore's Law advances that are exploitable every 18 months. The same technologies are actually subcomponents in the payload, communication, command system, and remote sensor support systems that currently compose various unmanned system of systems configurations.

Scenario: Warfighters need next-generation system capability to support an 18-month battlespace fielding requirement. The system involves new sensors supporting a remote sensor team. The platform will utilize several new algorithms to mitigate human support functions limited by human reaction time and communication anomalies. The platform will also have new decision algorithms supporting mission functions due to new payload capabilities. The system will use a new communication protocol evaluation system for onboard teams and for ground control teams communicating over a series of relay station and satcom grids. Platform support includes mission-driven T&E to validate autonomous support capabilities enabling nonlethal support functions. This migration will use technology for air traffic management that the Federal Aviation Administration (FAA) and DoD are co-evaluating for the automated notification of an aircraft's position to ground-based controllers as well as to other manned and unmanned aircraft.

The support situation calls for an assessment of the latest Standardization Agreement (STANAG) recovery algorithms in the event of communications link disruption with collaborating manned and unmanned systems supporting a teaming operation. Interoperability tests will be necessary to support several new services and remote service support teams translating data from the payload. The teams will leverage mission information from a variety of semantic databases across the Defense Grid to generate actionable intelligence. A new aspect of this deployment will involve the utilization of both trainers and users interacting with simulators to explore the adequacy of human systems integration algorithms to discover problems for algorithm refinement and problem discovery. Red team T&E technology will expand scenario assessment to provide forecasting without historical data using Bayesian probability models utilizing expert opinions. The T&E system will be designed to determine false positives, false negatives, dynamic limits, and integrity limits regarding mission effectiveness, suitability, survivability, and effectiveness. This wholesale advancement in T&E will result in a tenfold reduction in cycle time and cost.

3. Current State

Over the past decade, unmanned systems have played an increasing role in U.S. military operations. DoD uses a vast array of unmanned systems, from underwater to the upper regions of the atmosphere, from the size of a matchbox to the size of a Boeing 737.

These unmanned systems continue to prove their value in combat operations in Afghanistan, where military operations are planned and executed in extremely challenging environments. Adversaries are fighting using increasingly unconventional means, taking cover in the surrounding populations, and employing asymmetric tactics to achieve their objectives. In future conflicts, we must be prepared for these tactics as well as a range of other novel methods, including so-called "hybrid" and anti-access approaches to blunting U.S. power projection. Unmanned systems will be critical to U.S. operations in all domains across a range of conflicts, both because of capability and performance advantages, and the ability for unmanned systems to take greater risk.

As unmanned systems have proven their worth on the battlefield, DoD has allocated an increasing percentage of its budget to developing and acquiring these systems. Table 1 below reflects the budget request allocated to the three unmanned domains: air, ground, and maritime.

Although unmanned systems have experienced widespread growth in funding, current world economic conditions and DoD initiatives necessitate increased efforts and focus toward the acquisition of affordable and convergent systems. DoD must continue to support diverse mission sets and capabilities, but must focus on acquiring Joint and interoperable platforms, systems, software, architecture, payloads and sensors due to today's increasingly austere fiscal environment. In addition, the ability for commanders to take risks with unmanned vehicles depends significantly on their cost. In order to be expendable, which is often the intent of building an unmanned system, the vehicle must be low-cost. The importance of procuring common platforms with core C2 systems cannot be overstated as it will yield enormous collective benefits by reducing training costs, reducing supply chain diversity, improving availability, and offering a cost-effective procurement path by exploiting the benefits of scale and software/technology reuse.

> Eliminate redundancy within warfighter portfolios.
>
> *–Under Secretary of Defense Memorandum for Acquisition Professionals, Better Buying Power, September 2010*

The cost overruns, schedule slips, and sustainability issues of unmanned systems cannot go unnoticed or unanswered. Operational T&E is not sufficient for addressing budget, schedule, and sustainment issues in unmanned systems acquisition. WSARA 2009 guidance set the stage for leveraging developmental T&E as a key factor in T&E strategy to address Milestone A and B test challenges. Unmanned system T&E must not only consider physics effects but other areas that have an effect on algorithm development such as human factors, autonomous functionality, peering, collaboration, and autonomy-driven, red-team-based T&E limit testing. The goal to gradually reduce the degree of human control and decision making required for the unmanned portion of the force structure will mean that autonomous functionality will gradually increase and new ways to test this functionality will be required.

The need to maintain simplicity and overcome bureaucracy in unmanned system acquisition is an ongoing challenge. As these programs transition to acquisition programs, there is a unique opportunity to enable productive process and oversight appropriate to producing safe, suitable, survivable, and effective systems in a rapid acquisition framework.

There is a need to leverage OA and open interfaces to overcome the problems associated with proprietary robotic system architectures. Standards and interface specifications need to be established to achieve modularity, commonality, and interchangeability across payloads, control systems, video/audio interfaces, data, and communication links. Standardization will enhance competition, lower life-cycle costs, and provide warfighters with enhanced unmanned capabilities that enable commonality and joint interoperability on the battlefield.

Addressing factors inhibiting the growth of unmanned systems will provide more interoperability, more autonomy, better artificial intelligence, better communications, human systems integration, training standardization, more propulsion and power options, and better MUM teaming. These factors are addressed through the Joint Capability Integration and Development System (JCIDS) process.

Table 1. 2011 President's Budget for Unmanned Systems ($ Mil)

Unmanned Funding ($ Mil)							
Fiscal Year Defense Prog		**FY11**	**FY12**	**FY13**	**FY14**	**FY15**	**Total**
Air	RDTE	1,106.72	1,255.29	1,539.58	1,440.57	1,296.25	6,638.40
	PROC	3,351.90	2,936.93	3,040.41	3,362.95	3,389.03	16,081.21
	OM	1,596.74	1,631.38	1,469.49	1,577.65	1,825.45	8,100.71
Domain Total		6,055.36	5,823.59	6,049.48	6,381.17	6,510.72	30,820.32
Ground	RDTE	0.00	0.00	0.00	0.00	0.00	0.00
	PROC	20.03	26.25	24.07	7.66	0.00	78.01
	OM	207.06	233.58	237.50	241.50	245.96	1,165.60
Domain Total		227.09	259.83	261.57	249.16	245.96	1,243.61
Maritime	RDTE	29.69	62.92	65.72	48.60	47.26	254.19
	PROC	11.93	45.45	84.85	108.35	114.33	364.90
	OM	5.79	4.71	3.76	4.00	4.03	22.28
Domain Total		47.41	113.08	154.32	160.94	165.62	641.37
All Unmanned	RDTE	1,136.41	1,318.21	1,605.29	1,489.16	1,343.52	6,892.59
	PROC	3,383.86	3,008.63	3,149.32	3,478.96	3,503.36	16,524.12
	OM	1,809.59	1,869.67	1,710.75	1,823.15	2,075.44	9,288.59
Domain Total		6,329.86	6,196.50	6,465.36	6,791.27	6,922.31	32,705.30

3.1. Requirements Development and Systems Acquisition

There has been substantial growth in unmanned platforms of all sizes and shapes with a corresponding increase in payload numbers and capability. Many of these systems have been rapidly acquired and immediately fielded for warfighter use through the Joint Urgent Operational Needs (JUON) process. JUONs have successfully added significant capability to joint warfighting. While those unmanned systems were rapidly developed to meet the immediate needs of the warfighter in the short term, they have not undergone rigorous requirements review and joint coordination through the normal JCIDS process, to include systems interdependencies and interoperability. Further, their long term affordability, sustainability, and potential to contribute to long term enterprise-wide capability portfolios have not been fully considered. Consequently, they have not received due consideration in the context of broader joint capability areas (JCA) which provide structure and organization to Requirements Development.

DoD is moving toward revision of the JCIDS process which will streamline urgent and deliberate Capability Development to enable requisite timeliness in meeting warfighter needs, while giving important consideration to long term affordability and sustainability. JCIDS is a key supporting process for DoD acquisition and Planning, Programming, Budgeting, and Execution (PPBE) processes. It ensures the capabilities required by the warfighter are identified with their associated operational performance criteria in order to successfully execute the missions assigned. This process allows better understanding of the warfighting needs early in capability development and provides a more comprehensive set of valid prioritized requirements. The Department's acquisition arm can then focus on choosing options to meet well defined requirement capability.

Given today's highly constrained fiscal environment, it is imperative that the Department look at many areas where efficiencies can be gained to create unmanned systems that are both effective and affordable. The DoD will look at capitalizing upon commonality, standardization, and joint acquisition strategies among others. Also, the Department demands these unmanned systems be affordable at the outset and not experience significant cost growth in their development and production evolution. Additionally, it must provide the PPBE process with affordability advice by assessing the development and production lifecycle cost at the outset.

Capability requirements, validated by the JCIDS process, inform prioritization activities in the competition for funding during the PPBE process. The objective of the PPBE process is to provide the best mix of forces, equipment, and support attainable within fiscal constraints according to DoD Directive 7045.14, *Planning, Programming, Budgeting System (PPBS)*. To meet this objective, the PPBE process aims to meet goals established by the President and the Secretary of Defense (SECDEF) in the Strategic Planning and Joint Planning Guidance. In the PPBE process, the Services match available resources (fiscal, manpower, material) against validated requirements to achieve the strategic plan. A key task is to develop a balanced/affordable capabilities-based Service program objective memorandum (POM). The POM position for the capability to meet a given requirement is reviewed by OSD and the final position becomes the President's Budget.

The Joint Capability Areas (JCAs)[2] are currently the preferred method the Department of Defense uses for reviewing and managing capabilities. The JCA framework provides the structure around which capabilities and capability gaps can be aligned across the Department and across the various portfolios to correlate similar needs, leverage effective solutions, and synchronize related activities. Also, various frameworks, such as the Universal Joint Task List (UJTL), are readily available to aid in identifying and organizing the tasks, conditions and required capabilities.

3.2. Unmanned Systems Applied to Joint Capability Areas

Mapping current and projected unmanned systems against the JCAs provides a sense of the Product Line Portfolio of unmanned systems and how it currently, and could in the future, contribute to the missions of the Department. Each JCA represents a collection of related missions and tasks that are typically conducted to bring about the desired effects associated with that capability. Nine Tier One JCAs are defined, and assessments identified that unmanned systems have the potential to be key contributors for Battlespace Awareness, Force Application, Protection, Logistics, and Building Partnerships. Although assessments have not yet been completed for the Force Support and Net Centric capability areas, missions and tasks in those JCAs receive significant support from unmanned systems as well.

Current technology and future advancements can and will enable single platforms to perform a variety of missions across multiple capability areas. This represents an opportunity for the Department to achieve a greater return on investment. Furthermore, the projections show that there will be opportunities for joint systems to conduct missions for each of the Services, just as there will be situations in which domain conditions or Service missions will dictate unique solutions. Detailed descriptions of each of the systems identified for the capability areas, including specific tasks, performance attributes and integrated technologies can be found at the Unmanned Warfare Information Repository site: https://extranet.acq.osd.mil/uwir/. Below are the descriptions for the most relevant JCA.

3.2.1. Battlespace Awareness (BA)

Battlespace Awareness is a capability area in which unmanned systems in all domains have the ability to significantly contribute well into the future to conduct ISR and environment collection related tasks. To achieve this, unmanned systems development and fielding must include the Tasking, Production, Exploitation, and Dissemination (TPED) processes required to translate vast quantities of sensor data into a shared understanding of the environment. There are many ongoing efforts to streamline TPED processing. Applications in this JCA range from tasks such as aerial and urban reconnaissance, which is performed today by Predators, Reapers and Global Hawks in the air and by PackBots and Talons on the ground, to tasks such as Expeditionary Runway Evaluation, Nuclear Forensics, and Special Reconnaissance. In the future, technology will enable mission endurance to extend from hours to days to weeks so that unmanned systems can conduct long endurance persistent reconnaissance and surveillance in all domains. Because unmanned systems will progress further with respect to full autonomy, on-board sensors that provide the systems with their own organic perception will contribute to Battle Space Awareness regardless of their intended primary mission. This capability area is one that lends itself to tasks and missions being conducted collaboratively across domains, as well as teaming within a single domain.

3.2.2. Force Application (FA)

Force Application is another JCA which includes a proliferation of unmanned systems contributing to maneuver and engagement. Today, Predator, Reaper and Gray Eagle UAS are weaponized to conduct offensive operations, irregular warfare, and high value target / high value individual prosecution, and this trend will likely continue in all domains. In the air domain, projected mission areas for UAS include air-to-air combat and suppression and defeat of enemy air defense. On the ground, UGVs are projected to conduct missions such as non-lethal crowd control, dismounted offensive operations, and armed reconnaissance and assault operations. In the maritime domain, UUVs and USVs are projected to be particularly suited for mine laying and mine neutralization missions.

DoD personnel must comply with the law of war, including when using autonomous or unmanned weapon systems. For example, Paragraph 4.1 of DoD Directive 2311.01E, DoD Law of War Program, May 9, 2006, requires that: "[m]embers of the DoD Components comply with the law of war during all armed conflicts, however such conflicts are characterized, and in all other military operations." Current armed unmanned systems deploy lethal force only in a fully human-operated context (level 1) for engagement decisions. For these systems, the decisions both to employ force and to choose which specific target to engage are made by a human. The United States operates defensive systems for manned ships and installations that have human-supervised autonomous modes (level 3), and has operated these systems for decades. For the foreseeable future, decisions over the use of force and the choice of which individual targets to engage with lethal force will be retained under human control in unmanned systems.

3.2.3. Protection

Protection has particular unmanned systems applicability to assist in attack prevention or effects mitigation. Unmanned systems are ideally suited for many protection tasks that are deemed dull, dangerous or dirty. As the future enables greater automation with respect to both navigation and manipulation, unmanned systems will be able to perform tasks such as fire fighting, decontamination, forward operating base security, installation security, obstacle construction and breaching, vehicle and personnel search and inspection, mine clearance and neutralization, sophisticated explosive ordnance disposal, casualty extraction and evacuation, and maritime interdiction. In the Protection JCA teaming within domains and collaboration across domains will likely prevail.

3.2.4. Logistics

The Logistics joint capability area is also ideally suited for employing unmanned systems in all domains to deploy, distribute, and supply forces. Transportation of supplies is an applicable, routine task, particularly suited for unmanned systems in all types of ground terrain.

Maintenance related tasks such as inspection, decontamination, and refueling can be performed by unmanned systems. Munitions and material handling, and combat engineering are ideal tasks that can be allocated to unmanned systems to enhance safety as well as increase efficiency. Additionally, casualty evacuation and care, human remains evacuation, and urban rescue can also be tasks performed by unmanned systems. Unmanned systems will perform Logistics tasks on home station as well as forward deployed.

Table 2. DoD Unmanned Capabilities by Program below is a sample mapping of JCA tasks to the current unmanned inventory and is provided for determining current unmanned systems capabilities.

3.3. Unmanned Aircraft Systems (UAS)

The air domain has received the greatest concentration of visibility as DoD has embraced unmanned technologies. Table 1 depicts that UAS investments will continue to consume a large share of the overall DoD investment in unmanned systems. These efforts have fielded a large number of UAS capable of executing a wide range of missions. Originally, UAS missions focused primarily on tactical reconnaissance; however, this scope has been expanded to include most of the capabilities within the ISR and battlespace awareness mission areas. UAS are also playing a greater role in strike missions as the military departments field multiple strike mission-capable weapon systems for time-critical targeting. Figure 1 below illustrates the variety of platforms in today's force structure.

As the number of fielded systems has expanded, flight hours have dramatically increased as depicted in Figure 2.

In 2009, DoD completed almost 500,000 UAS flight hours just in support of Operation Enduring Freedom and Operation Iraqi Freedom. In May 2010, unmanned systems surpassed one million flight hours and in November 2010 achieved one million combat hours. As these systems continue to demonstrate their value, this number will continue to grow.... remotely piloted vehicles have flown more than 21,000 sorties so far this year, already surpassing the roughly 19,000 drone flights last year.

– *"U.S. Uses Attacks to Nudge Taliban Toward a Deal," New York Times, October 15, 2010.*

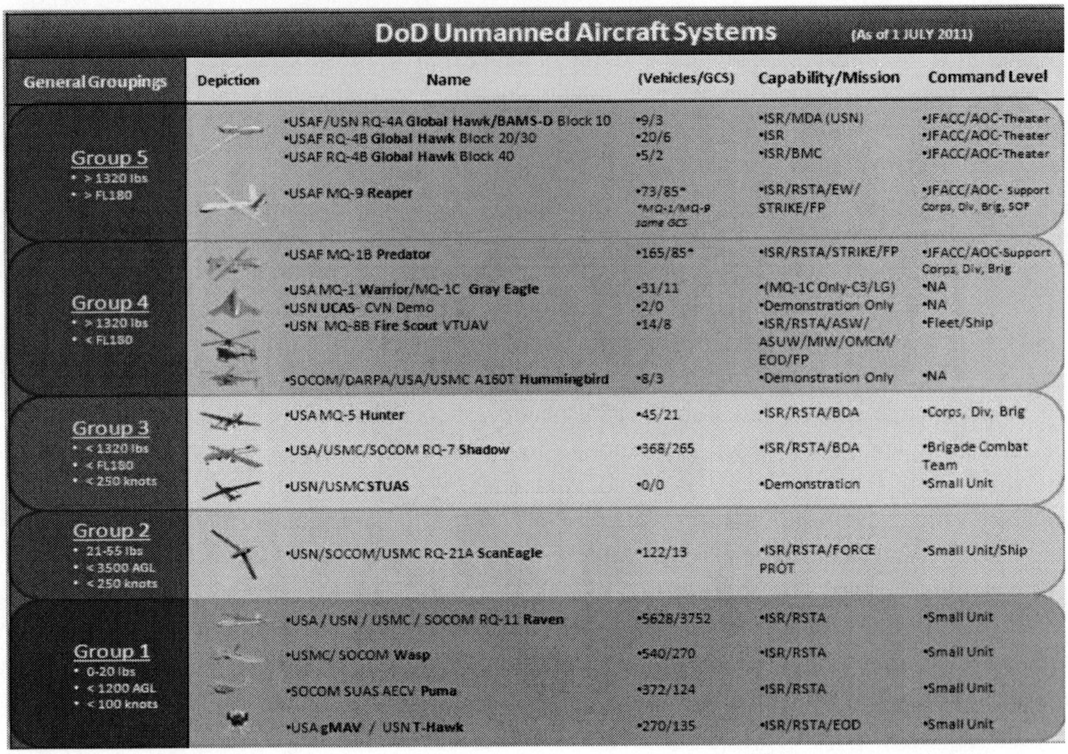

DoD Unmanned Aircraft Systems (As of 1 JULY 2011)

General Groupings	Depiction	Name	(Vehicles/GCS)	Capability/Mission	Command Level
Group 5 • > 1320 lbs • > FL180		•USAF/USN RQ-4A **Global Hawk/BAMS-D** Block 10 •USAF RQ-4B **Global Hawk** Block 20/30 •USAF RQ-4B **Global Hawk** Block 40	•9/3 •20/6 •5/2	•ISR/MDA (USN) •ISR •ISR/BMC	•JFACC/AOC-Theater •JFACC/AOC-Theater •JFACC/AOC-Theater
		•USAF MQ-9 **Reaper**	•73/85* *MQ-1/MQ-9 same GCS	•ISR/RSTA/EW/ STRIKE/FP	•JFACC/AOC- Support Corps, Div, Brig, SOF
Group 4 • > 1320 lbs • < FL180		•USAF MQ-1B **Predator**	•165/85*	•ISR/RSTA/STRIKE/FP	•JFACC/AOC-Support Corps, Div, Brig
		•USA MQ-1 **Warrior**/MQ-1C **Gray Eagle** •USN **UCAS**- CVN Demo •USN MQ-8B **Fire Scout** VTUAV	•31/11 •2/0 •14/8	•(MQ-1C Only-C3/LG) •Demonstration Only •ISR/RSTA/ASW/ ASUW/MIW/OMCM/ EOD/FP	•NA •NA •Fleet/Ship
		•SOCOM/DARPA/USA/USMC A160T **Hummingbird**	•8/3	•Demonstration Only	•NA
Group 3 • < 1320 lbs • < FL180 • < 250 knots		•USA MQ-5 **Hunter**	•45/21	•ISR/RSTA/BDA	•Corps, Div, Brig
		•USA/USMC/SOCOM RQ-7 **Shadow**	•368/265	•ISR/RSTA/BDA	•Brigade Combat Team
		•USN/USMC **STUAS**	•0/0	•Demonstration	•Small Unit
Group 2 • 21-55 lbs • < 3500 AGL • < 250 knots		•USN/SOCOM/USMC RQ-21A **ScanEagle**	•122/13	•ISR/RSTA/FORCE PROT	•Small Unit/Ship
Group 1 • 0-20 lbs • < 1200 AGL • < 100 knots		•USA / USN / USMC / SOCOM RQ-11 **Raven**	•5628/3752	•ISR/RSTA	•Small Unit
		•USMC/ SOCOM **Wasp**	•540/270	•ISR/RSTA	•Small Unit
		•SOCOM SUAS AECV **Puma**	•372/124	•ISR/RSTA	•Small Unit
		•USA **gMAV** / USN **T-Hawk**	•270/135	•ISR/RSTA/EOD	•Small Unit

Figure 1. DoD UAS.

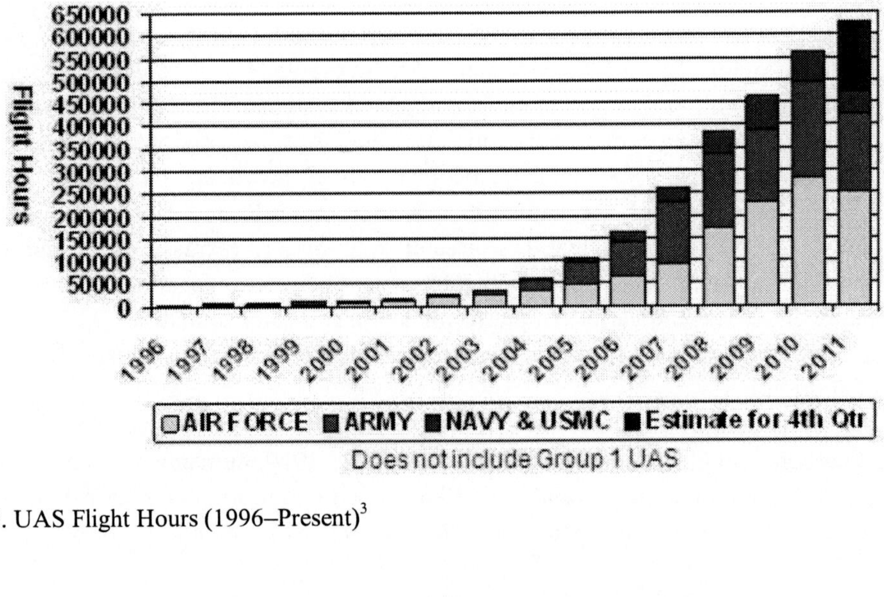

Figure 2. UAS Flight Hours (1996–Present)[3]

Table 2. DoD Unmanned Capabilities by Program

System	Lead Service	Primary JCA	Mission Capabilities	ACAT	Acquisition Status
AIRCRAFT					
GROUP 1					
RQ-16B T_Hawk	US Navy	N/A	ISR/RSTA, EOD	Non-ACAT	Other
Wasp	US Air Force	BA	ISR/RSTA	Non-ACAT	Other
RQ-11B Raven	US Army	BA	ISR/RSTA	IV(T)	Production
Puma AE	USSOCOM	N/A	ISR/RSTA, FP	III	Production/Sustainment
GROUP 2					
Scan Eagle	US Navy, US Marines	N/A	ISR/RSTA, Force Protection	Non-ACAT	Other
GROUP 3					
RQ-7B Shadow	US Army, US Marines	BA	ISR/RSTA, C3, Force Protection	II	Production
S 100	USSOCOM	N/A	ISR/RSTA, EW, Force Protection	III	Design & Development
STUAS RQ-21A	US Navy, US Marines	BA	ISR/RSTA, EOD, Force Protection	III	Design & Development
Viking 400	USSOCOM	N/A	ISR/RSTA, EW, Force Protection	III	Design & Development
GROUP 4					
MQ-5B Hunter	US Army	N/A	ISR/RSTA, C3, Log, PS/TCS, FP	N/A	Other
MQ-1C Gray Eagle	US Army	BA	ISR/RSTA, C3, Log, PS/TCS, FP	I D	Production
MQ-1B Predator	US Air Force	BA	ISR/RSTA, PS/TCS, FP	I D	Sustainment
MQ-8B VTUAV	US Navy		ISR/RSTA, ASW, SUW/ASUW,	I C	MS-C
GROUP 5					
MQ-4 BAMS	US Navy		ISR/RSTA, EW, PS/TCS, SUW/ASUW, FP	I D	Design &Development
MQ-9A Reaper	US Air Force	FA	ISR/RSTA, EW, PS/TCS, FP	I D	Production
RQ-4A Global Hawk	US Air Force	BA	ISR/RSTA, C3, PS/TCS	I D	Sustainment
RQ-4B Global Hawk	US Air Force	BA	ISR/RSTA, C3, PS/TCS	I D	Production/Sustainment
MR UAS	US Navy	N/A	TBD	N/A	Concept
UCLASS	US Navy	N/A	TBD	N/A	Concept
MQ-X	US Air Force	FA	ISR/RSTA, PS/TCS, FP	N/A	Concept
Group 4	US Marines	N/A	TBD	N/A	Concept
GROUND VEHICLES					
WHEEL					
MARCbot IV N	US Army	N/A	ISR/RSTA, IED Inv.	Other	Other

Table 2. (Continued)

System	Lead Service	Primary JCA	Mission Capabilities	ACAT	Acquisition Status
GROUND VEHICLES					
Throwbot	US Army	N/A	ISR/RSTA	Other	Other
Mine Area Clearance Equipment (MACE)	US Air Force	N/A	Mine, EOD, FP	Other	Concept
Defender	US Air Force	N/A	FA, FP, Fire	Other	Concept
TRACK					
ISR UGV	US Navy	N/A	ISR/RSTA, Fire Support, EOD	Other	Other
xBot	US Army	N/A	ISR/RSTA, EOD, IED Inv.	Other	Other
PackBot FIDO	US Army	N/A	ISR/RSTA, EOD, IED Inv.	Other	Other
M 160	US Army	N/A	Mine Neutralization	III	Design &Development
RC50 60	US Army	N/A	EOD, Mine Neutralization	Other	Other
Mini EOD	US Army	N/A	EOD	Other	Other
ANDROS HD 1	US Army	N/A	EOD	Other	Other
PackBot EOD	US Army	N/A	EOD	Other	Other
TALON IIIB	US Army	N/A	EOD, route clearance	Other	Other
TALON IV	US Army	N/A	EOD, route clearance	Other	Other
Panther II	US Army	N/A	EOD, Mine Neutralization	Other	Other
MK 1 MOD 0 Robot EOD	US Navy	N/A	EOD	IV	Sustainment
MK 2 MOD 0 Robot EOD	US Navy	N/A	EOD	IV	Sustainment
MK 2 MOD 0 RONS	US Navy	N/A	EOD	IV	Sustainment
All-Purpose Remote Transport System (ARTS)	US Air Force	N/A	Mine, EOD, FP, Fire	Other	Other
F6A ANDROS	US Air Force	N/A	EOD	Other	Other
HD 1	US Air Force	N/A	EOD	Other	Other
IVAN	US Air Force	N/A	EOD, FP	Other	Concept
MARITIME CRAFT					
SURFACE					
Autonomous Unmanned Surface Vehicle (AUSV)	US Navy	N/A	ISR/RSTA	Other	Other
Mine Countermeasures (MCM) Unmanned Surface Vehicle USV	US Navy	BA	MIW/OMCM	Other	Concept
Anti-Submarine Warfare (ASW) Unmanned	US Navy	N/A	ASW	Other	Other
Surface Vehicle (USV)					

MARITIME CRAFT					
System	**Lead Service**	**Primary JCA**	**Mission Capabilities**	**ACAT**	**Acquisition Status**
Sea Fox	US Navy	N/A	ISR/RSTA, FP	Other	Other
Remote Minehunting System (RMS), AN/WLD-1(V)1	US Navy	BA	MIW/OMCM	I D	Design & Development
Modular Unmanned Surface Craft Littoral	US Navy	N/A	ISR/RSTA	Other	Other
UNDERWATER					
Sea Stalker	US Navy	N/A	ISR/RSTA	Other	Other
Sea Maverick	US Navy	N/A	ISR/RSTA	Other	Other
Echo Ranger	Commercial	N/A	Insp/ID, Oceanogrpahic Survey	Other	Other
Marlin	Commercial	N/A	Insp/ID, Oceanogrpahic Survey	Other	Other
Surface Countermeasure Unmanned Undersea Vehicle	US Navy	BA	MIW/OMCM	III	Concept
MK18 Mod 2 Kingfish UUV System	US Navy	Protection	SUW/ASUW, MIW/OMCM, Insp/ID	PIP	Production
Surface Mine Countermeasure Unmanned Undersea	US Navy	N/A	MIW/OMCM	Other	Other
Vehicle User Operational Evaluation System Increment 2	US Navy	N/A	MIW/OMCM	Other	Other
Vehicle User Operational Evaluation System Increment 1	US Navy	N/A	MIW/OMCM	Other	Other
Battlespace Preparation Autonomous Underwater Vehicle (BPAUV)	US Navy	N/A	MIW/OMCM	Other	Other
HULS	US Navy	Protection	MIW/OMCM, EOD, Insp/ID	Abbr Acq	Production
MK18 Mod 1 Swordfish UUV System	US Navy	Protection	MIW/OMCM, EOD, Insp/ID	Abbr Acq	Sustainment
Large Displacement Unmanned Underwater Vehicle (LDUUV)	US Navy		ASW, ISR, MCM	Other	Concept
MK18 Mod 1 Swordfish UUV System	US Navy		MIW/OMCM, EOD, Insp/ID	Abbr Acq	Sustainment

3.4. Unmanned Ground Systems (UGS)

Since operations in Iraq and Afghanistan began, DoD has acquired and deployed thousands of UGS. These systems support a diverse range of operations including maneuver, maneuver support, and sustainment. Maneuver operations include closing with and neutralizing the enemy using speed and firepower. Maneuver support missions include mitigating natural and artificial obstacles and hazards. Sustainment missions leverage maintenance and support UGVs associated with combat services support.

Approximately 8,000 UGVs of various types have seen action in Operation Enduring Freedom and Operation Iraqi Freedom. As of September 2010, these deployed UGVs have been used in over 125,000 missions, including suspected object identification and route clearance, to locate and defuse improvised explosive devices (IEDs). During these counter-IED missions, Army, Navy, and USMC explosive ordnance teams detected and defeated over 11,000 IEDs using UGVs, such as the one depicted in Figure 3.

Figure 3. Talon Ordnance Disposal Robot Preparing to Unearth Simulated IED.

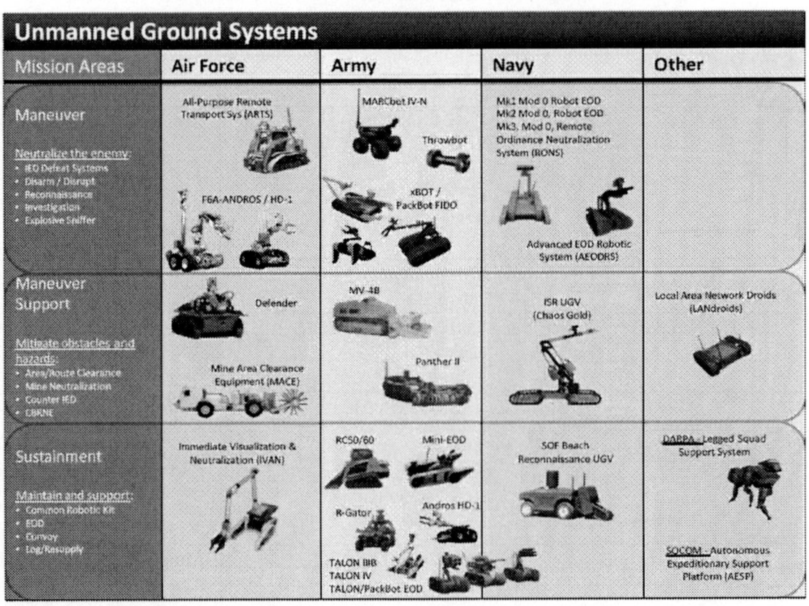

Figure 4. UGS FoS.

The lessons collected on the battlefield must be translated into programs that can be sustained. The rapid fielding and proliferation of unmanned systems and the subsequent battlefield modernization they provided have met the mission, but resulted in configuration and maintenance challenges. These ground systems continue to provide tremendous benefit to the ground commander, but improvements in user interfaces, reliability, survivability, and advances in 360° sensing, recording fidelity, and CBRN and explosive detection are required to meet the challenges anticipated in future conflicts. Figure 4 shows the UGS Family of Systems (FoS).

> On Gordon's (UGV) final days, he was launched out of the truck and was searching an intersection for a possible deep buried IED. As he was on his way to the intersection, the IED was detonated about 10 ft from his location. Still functioning, he continued to search the area. On the opposite side of the road, another IED was detonated and had turned him upside down. Everything was still working until a fire fight started.
>
> Gordon took 7 rounds to the underside and was done for the day. I took him to the robot shop for repair. It took 3 days. When he was returned to us, I put him back in the truck to get him back on duty. But this was shortly lived as he was searching a gate at a house for possible booby-traps that detonated directly next to him. Gordon was mangled beyond repair. Now his replacement "Flash" is here to finish his job.
>
> -- Insight from an End User: *"Gordon" TALON Defeats IEDs and Saves Lives in Baghdad*, submitted by an EOD operator, summer 2007, Iraq.

3.5. Unmanned Maritime Systems (UMS)

Over 90% of the information, people, goods, and services that sustain and create opportunities for regional economic prosperity flow across the maritime domain. With emerging threats such as piracy, natural resource disputes, drug trafficking, and weapons proliferation, a rapid response capability is needed in all maritime regions. DoD continues to expand the range of missions supported by unmanned systems in the maritime domain. A recent study concluded that unmanned maritime systems "have the potential to provide critical enabling capabilities for current NATO [North Atlantic Treaty Organization] maritime missions that can improve Alliance security and stability".[4]

> USVs, along with UUVs, will have an important role in the conduct of MCM [mine countermeasures] as they are particularly well suited for the 'dirty - dull - dangerous' tasks that MCM entails. They provide persistence, which permits significant mine hunting and sweeping coverage at lower cost by multiplying the effectiveness of supporting or dedicated platforms. Additionally, they provide the potential for supporting an MCM capability on platforms not traditionally assigned a mine warfare mission.
>
> *– USV mission descriptions, The Navy Unmanned Surface Vehicle Master Plan,23 July 2007*

Like UAS and UGS, UMS have the potential to save lives, reduce human risk, provide persistent surveillance, and reduce operating costs. UMS priority missions are listed below.

UMS can be defined as unmanned vehicles that displace water at rest and can be categorized into two subcategories: unmanned underwater vehicles (UUV) and unmanned surface vehicles (USV). USVs are UMS that operate with near-continuous contact with the surface of the water, including conventional hull crafts, hydrofoils, and semi-submersibles.[5] UUVs are made to operate without necessary contact with the surface (but may need to be near surface for communications purposes) and some can operate covertly.

The use of UMS is not new. After World War II, USVs were used to conduct minesweeping missions and test the radioactivity of water after each atomic bomb test. Another example occurred during the Vietnam War in an area south of Saigon, where remotely controlled USVs conducted minesweeping operations. More recently, UUVs conducted mine-clearing activities during Operation Iraqi Freedom in 2003. A complementary suite of UMS serve as the foundation for MCM operations from the Littoral Combat Ship (LCS) and small diameter UUVs are currently the main mine detection capability for ports & harbor and in the Very Shallow Water zone.

At a recent Science and Technology Conference hosted by the Office of Naval Research, Chief of Naval Operations (CNO) Admiral Gary Roughead made a number of statements expressing UMS goals for the Department of the Navy. Solving the power consumption problem would be the "one thing" CNO would most like to see the Navy's scientists accomplish.[6] Rear Admiral Nevin Carr, Chief of Naval Research, explained the current efforts in filling the needs in this unmanned maritime area and went on to describe where and how this technology might be applied in the future: "Two options are under exploration: fuel-cell technologies and radioscope thermoelectric generators that can provide low amounts of power for very long periods of time. We might start thinking about setting up drone refueling stations. You might deploy a remotely manned underwater generator that sits on the bottom in a secure area, which is a secure location where your forward-deployed vehicles might come back and recharge."

UMS Priority Missions

Mine Countermeasures (MCM)
Anti-Submarine Warfare (ASW)
Maritime Domain Awareness
Maritime Security

Figure 5 illustrates the variety of platforms and maritime missions supporting today's operations by UMS and those planned for operation in the near future. They have the potential for even greater integration, especially UUVs, to the point of assisting the current submarine and surface fleet in replacing fixed underwater sensor grids; using UUVs and distributed netted sensors to expand our submarine's sphere of influence; and weaponizing UUVs.

However, there are currently limitations to realizing the full potential of UMS:

- Endurance
- Underwater C2 and deconfliction

- Survivability in an unforgiving environment
- Launch and recovery
- Communication technology for dynamic tasking, querying, and data dissemination

These challenges are areas for further technical exploration. Despite these challenges, the future for UMS is very promising. Building on the experience and contribution from this first generation of fielded UMS, the shift is underway from UMS merely serving as an extension of the sensor systems of manned ships and submarines into an integrated FoS to provide full mission capabilities with increased autonomy.

CHALLENGES

Affordability

Interoperability
Autonomy
Airspace Integration
Communications
Training
Propulsion and Power
Manned-unmanned (MUM) Teaming

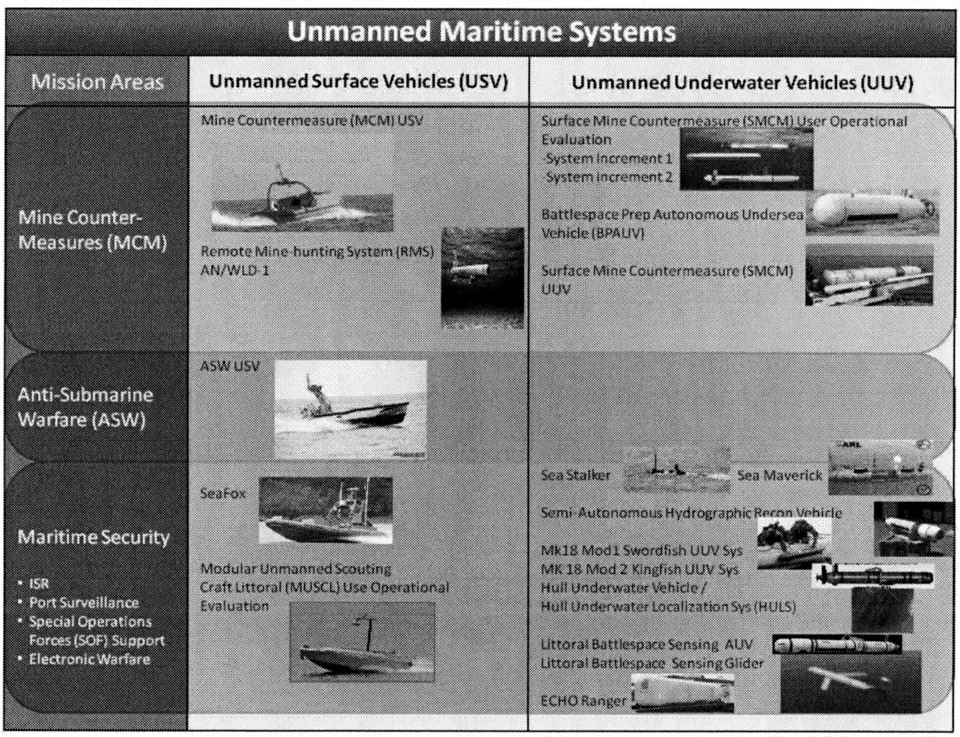

Figure 5. DoD UMS FoS.

3.6. Challenges for Unmanned Systems

The number of fielded systems and the range of missions supported by unmanned systems continue to grow at a dramatic rate. As DoD steers a path toward the vision described in Section 2, the challenges listed on the right must be overcome in order to realize the full potential offered by unmanned systems. The following subsections summarize these challenges and the remainder of this document provides details and future goals for dealing with each challenge.

3.6.1. Interoperability

To maximize the potential of unmanned systems, the systems must be capable of operating seamlessly with each other and with manned systems across the air, ground, and maritime domains. System interoperability is critical in achieving these objectives and requires the implementation of mandated standards and Interoperability Integrated Product Team (I-IPT) profiles. Properly implemented, interoperability can serve as a force multiplier, improve joint warfighting capabilities, decrease integration timelines, simplify logistics, and reduce total ownership costs (TOC). One of the most powerful tools in maximizing interoperability and achieving these objectives is the adoption of the open systems architecture concept.

3.6.2. Autonomy

The rapid proliferation of unmanned systems and the simultaneous operation of manned and unmanned systems as unmanned systems expand into additional roles have created a manpower burden on the Services. With limited manpower resources to draw upon, the Services are seeking ways to improve the efficiency of operations. For instance, introducing a greater degree of system autonomy will better enable one operator to control more than one unmanned system, and has the potential to significantly reduce the manpower burden. Additional benefits are greatly reducing high bandwidth communication needs and decreasing decision cycle time. Similar efficiencies can be gained by automating the tasking, processing, exploitation, and distribution (TPED) of data collected by unmanned systems. Autonomy can help extend vehicle endurance by intelligently responding to the surrounding environmental conditions (e.g., exploit/avoid currents) and appropriately managing onboard sensors and processing (e.g., turn off sensors when not needed). Implementing a higher degree of autonomy faces the following challenges:

- Investment in science and technology (S&T) to enable more capable autonomous operations
- Development of policies and guidelines on what decisions can be safely and ethically delegated and under what conditions
- Development of new Verification and Validation (V&V) and T&E techniques to enable verifiable "trust" in autonomy

3.6.3. Airspace Integration (AI)

The rapid increase in fielded UAS has created a strong demand for access within the NAS and international airspace. The demand for airspace to test new systems and train UAS operators has quickly exceeded the current airspace available for military operations. Figure 6 shows the projected number of DoD UAS locations in the next six years, many without access to airspace compatible for military operations under the current regulatory environment.

NAS access for UAS is currently limited primarily due to regulatory compliance issues and interim policies. DoD UAS operations conducted outside of restricted, warning, and prohibited areas are authorized only under a (temporary) Certificate of Waiver or Authorization (COA) from the FAA. The COA process is adequate for enabling a small number of flights, but does not provide the level of airspace access necessary to accomplish the wide range of DoD UAS missions at current and projected operational tempos (OPTEMPOs). This constraint will only be exacerbated as combat operations in Southwest Asia wind down and systems are returned to U.S. locations.

3.6.4. Communications

Current unmanned systems operations involve a high degree of human interaction with the systems via various means for C2 and transmission of operational data. Protection of these communication links and the information flowing through them is critical to these operations. As the number of fielded systems grows, communications planners face challenges such as communication link security, radio frequency spectrum availability, deconfliction of frequencies and bandwidth, network infrastructure, and link ranges. Intelligent means of data parsing is needed to enable TPED and counter communication challenges.

3.6.5. Training

The rapid proliferation of numbers and types of UAS in response to wartime demand coupled with an expected redeployment of forces back to a peacetime footing will result in continuation training greater than what has been required in the past. This has caused pause to examine UAS training and to develop an overall strategy. At present, due to high demand for UAS assets in real world contingencies, most day-to-day, continuation training is accomplished under in-theater combat conditions. At the same time, disparate efforts by a number of organizations across the Department are underway to try to address UAS training requirements. As UAS forces drawdown in theater and redeploy, the Services will require comprehensive continuation and Joint force training in the peacetime environment at UAS bed-down and selected Joint training locations.

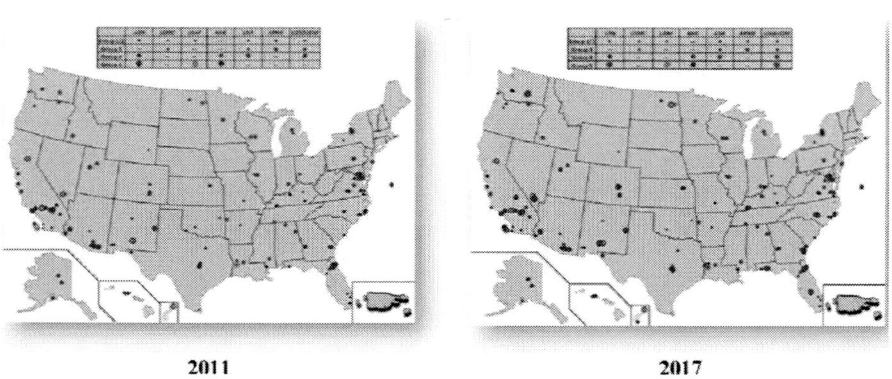

Figure 6. Representative DoD UAS Locations from 2011 to 2017.

3.6.6. Propulsion and Power

The dramatic increase in the development and deployment of unmanned systems across the entire spectrum of air, ground, and maritime missions has led to a concurrent increase in the demand for efficient, powerful, often portable, and logistically supportable solutions for unmanned system propulsion and power plant requirements. As these systems continue to demonstrate their value, operators want them to function longer without refueling and to do more tasks; these demands tax the internal power sources. The laboratories of the military Services and industry are focusing their efforts to find efficient solutions to the demand for improved propulsion and power plants. Regardless of energy source, total vehicle design, from materials used to autonomous response to the physical environment, needs to be considered up front to maximize endurance.

3.6.7. Manned-Unmanned (MUM) Teaming

MUM teaming refers to the relationship established between manned and unmanned systems executing a common mission as an integrated team. U.S. military forces have demonstrated early progress in integrating unmanned systems within the existing manned force structure, but much more needs to be done to achieve the full potential offered by unmanned technology. Improving MUM teaming is both a technology challenge (such as connecting the systems) and a policy challenge (such as establishing the rules of engagement for operating semi-autonomous unmanned with manned systems).

4. Interoperability

4.1. Overview

There is a clear benefit for warfighters to be able to seamlessly command, control, communicate with, exploit and share sensor information from unmanned systems across multiple domains. The Unmanned Systems Interoperability Initiative (UI2) led by the OSD UAS Task Force is in the process of developing an overarching strategy for increasing unmanned systems interoperability, with the long-range vision of producing a strategy that can be leveraged across the full spectrum of both unmanned and manned systems. DoD's goal is to move from Service/Agency-unique, stand-alone capabilities toward substantially improved interoperability standards that lead to an improved collaborative operational environment.

> Lack of UAV interoperability has had a real-life impact on U.S. operations … there have been cases where a Service's UAV, if it could have gotten data to another Service, another component, it may have provided better situational awareness on a specific threat in a specific area that might have resulted in different measures being taken.
>
> – *Dyke Weatherington (PSA/UW)*

4.2. Functional Description

Interoperability is the ability to operate in synergy in the execution of assigned tasks.[7] Properly implemented, it can serve as a force multiplier, improve warfighter capabilities, decrease integration timelines, simplify logistics, and reduce TOC. DoD Directive (DODD) 5000.1 establishes the requirement to acquire systems and FoSs that are interoperable.[8] DoD's unmanned systems will need to demonstrate interoperability in a number of areas:

- *Among similar components of the same or different systems.* The plug-and-play use of different sensors on an unmanned vehicle.
- *Among different systems of the same modality.* An open common ground control station (GCS) architecture for multiple, heterogeneous unmanned vehicles.
- *Among systems of different modalities.* The ability of air, ground, and maritime vehicles to work cooperatively.
- *Among systems operated by different Military Departments under various CONOPS and TTP, i.e., in joint operations.* Joint service systems working in concert to execute a common task or mission.
- *Among systems operated and employed by coalition and allied militaries under the governance of various concepts of employment (CONEMPs), TTPs, e.g., in multinational combined operations or NATO STANAGs.* The ability of coalition and allied systems to work in concert to execute a common task or mission based on predefined roles and responsibilities.
- *Among military systems and systems operated by other entities in a common environment.* The ability of military UAS to share the NAS and international airspace with commercial airliners and general aviation.
- *Among systems operated by non-DoD organizations, Allies, and coalition partners, i.e., in combined operations.* The ability of assets from organizations such as Customs and Border Protection (CBP) and Department of Homeland Security (DHS) to coordinate, interoperate, and exchange information with DoD assets of the same modality and same model.

> The interoperability goal for Unmanned Systems is an ability to provide data, information, material, and services to and accept the same from other systems, units, or forces … and to use the exchanged data, information, material, and services to enable them to operate effectively together.

The Joint Unmanned Aircraft Systems Center of Excellence (JUAS COE)[9] maintains the Joint CONOPS for UAS, which provides a joint vision for the operation, integration, and interoperability of UAS and touches on several of the areas mentioned above. Figure 7 illustrates joint, cross-domain interoperability.

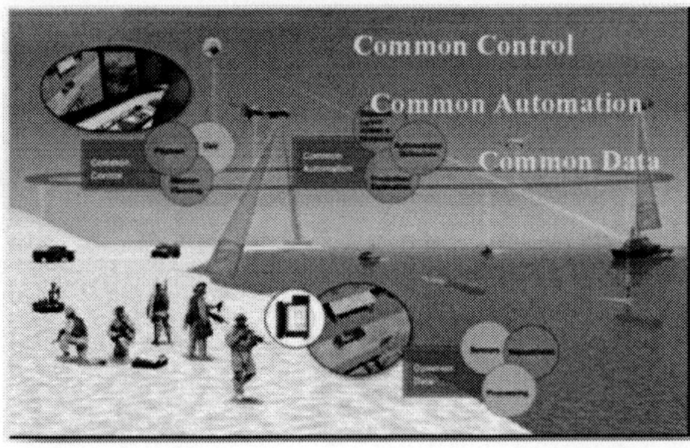

Figure 7. Joint Cross-Domain Interoperability.

4.3. Today's State

The historical approach to software and hardware acquisition relied on dedicated design for each system to accomplish a specific mission or capability. This approach may be optimal for a single system, but it unfortunately produces a collection of discrete, disjointed solutions with significant functional overlap and no method to exploit common components of each system.

Open architecture (OA) facilitates interoperability between systems by effectively leveraging the following concepts:

- Common capability descriptions in system requirements
- Common, open data models, standards, interfaces, and architectures in system design
- Common components in system acquisition strategies

OSD defines OA as a multifaceted strategy providing a framework for developing joint interoperable systems that adapt and exploit open-system design principles and architectures. This framework includes a set of principles, processes, and best practices that:

- Provide more opportunities for competition and innovation
- Rapidly field affordable, interoperable systems
- Minimize total ownership cost
- Optimize total system performance
- Yield systems that are easily developed and upgradeable
- Achieve component software reuse[10]

These (predominantly) acquisition issues are aided by a solid framework for software, component, and systems interoperability.

Traditionally, efforts have focused on system functionality descriptions, with interoperability focused at the messaging layer (e.g., the Joint Architecture for Unmanned Systems (JAUS) and STANAG 4586) to achieve standards-based interoperability. However, the tenets of common definitions and understanding listed on the left are required to achieve a true plug-and-play level of interoperability in which software capabilities from multiple vendors can be developed and integrated into a single system, supporting the exchange, interpretation, and action on data from other systems.

Through implementation and program-level adoption of these three tenets, DoD intends to address the issue of single proprietary vendor dependency within the acquisition system and improve conditions allowing for full competition.

TENETS OF COMMON DEFINITIONS FOR PLUG-AND-PLAY INTEROPERABILITY

System Functionally Descriptions and Architectures
Messaging Standards (e.g. STANAG 4586, JUAS, USMTF)
Data Models, including formats, definitions, semantics, texonomies, etc.

4.4. Problem Statement

Over the last decade, the DoD has achieved great successes from the use of unmanned systems for both peacetime and wartime operations. These successes led to a significant increase in the number of unmanned systems planned and procured, and the future looks to see an exponential increase in the quantity and diversity of unmanned systems applications. Traditionally, each unmanned system was procured as a vertically integrated, vendor-proprietary solution, consisting of the vehicle system, control station, communications channels, and encryption technologies. These single-system variants were typically "closed" systems utilizing proprietary interfaces. Development of the entire system was conducted in parallel with close interdependencies between components and procured as a whole through the platform prime contractor. As the number of new unmanned systems programmed in the Service budgets increased, the magnitude of RDT&E requirements for development skyrocketed. In addition to cost, this approach resulted in a number of unfavorable acquisition and growth characteristics that impeded progress as depicted above. Further, silence about the lack of interoperability and standards failed to foster dialog on how to overcome them. Over time, this resulted in an inhibition to innovation; increased vulnerability to threats where attacks on common attributes can impact multiple systems; increased complexity to systems engineering, development, and test; increased upfront costs; increased costs to system upgrades that cannot be made without changes to the interoperable dependencies of multiple systems; and budgeting protocols that treat interservice and coalition interoperability as fiscal trade-space.

UNFAVORABLE CHARACTERISTICS OF CURRENT APPROACH

- Lack of re-usability, resulting in RDT&E dollars being inefficiently utilized on repeated development of similar technology for different platforms.
- Difficulty of upgrading and enhancing capability due to the proprietary nature of UxVs.

- Inability to leverage research and development conducted at small businesses.
- Reliance on large prime vendors and vertical integrators who have little motivation for controlling and managing schedule.

These issues have significantly hampered unmanned systems acquisition activities. However, urgent wartime needs dictated that such concerns be relegated to the background, in the interest of rapid initial deployment. As the unmanned systems industry matures, however, the acquisition process must evolve in parallel. Addressing and enabling interoperability within unmanned systems will help accomplish this goal.

4.5. The Way Ahead

The technical approach to achieve the interoperability vision leverages the tight connection between interoperability and OAs, and consists of several elements. Each of the following elements is required, and none is sufficient in its own right to implement an OA:

- Development of a standard data model and service definitions that support OA concepts.
- Development of multiple repositories of models, software components, interface standards, and infrastructure services that can be used across the Services to extend, adapt, and compose unmanned systems and support software component reuse. These repositories should encourage the use of commercial, off-the-shelf (COTS) solutions where available, and are not intended to be "single point bottlenecks" as other efforts have been in the past. The goal is to provide multiple collection points across the Services for best practices, interfaces, and implementations.
- Collaboration among Government, industry, and academia to extend and manage the repositories and to validate components.
- Migration of current and developing systems to the OA approach.

To meet current interoperability standards, DoD will rely more heavily on spiral and incremental development initiatives ensuring services are compliant with these standards.

4.5.1. Open Architecture (OA)

OA utilizes a common set of interfaces and services; associated data models; robust, standard data busses; and methods for sharing information to facilitate development. OA involves the use of COTS components with published, standard interfaces, where feasible, at all levels of system design. This approach avoids proprietary, stove-piped solutions that are vendor-specific and enables innovation to be better captured and integrated into systems design. The OA approach allows for expanded market opportunities, simplified testing and integration, and enhanced reusability throughout the program life cycle. The Navy's Cruiser Modernization Program is one such effort.

The OA process encourages innovation, allows information sharing among competitors, and rewards Government and industry for this collaboration. It allows programs to include small businesses in systems acquisition activities as a valuable, affordable, and innovative source of technologies and capabilities. The result is a better product.

DoD unmanned systems consist of a wide range of programs, architectures, and acquisition approaches. To create a common framework for development and acquisition, DoD adopted principles of OA and service-oriented architecture (SOA). While the OA is the contracting, architecture, and business process methodology used to develop and acquire systems, a SOA is a specific way of designing software, in a standardized architecture, that uses interchangeable and interoperable software components called *services*. When coupled together, the result is a business approach to acquiring software developed within a common engineering construct that promotes reuse, cost reduction, competition, growth opportunity, expandability, innovation, and interoperability among similar systems.

SOA provides a set of principles or governing concepts that are used during the phases of systems development and integration. This type of architecture attempts to package functionality as interoperable services within the context of the various business domains that use it. SOAs increase functionality by incorporating new services, which are developed separately but integrated within the system's common framework as a new capability. Their interfaces are independent of application behavior and business logic, and this independence makes the interfaces agile in supporting application changes and enables operations across heterogeneous software and hardware environments.

Programs and efforts to date have strongly tied together unmanned systems capability requirements and definitions, along with underlying technology selections. In recognition of the rapidly changing technology, unmanned systems architectures would benefit strongly from being defined at a platform-independent model (PIM) level, which is devoid of technology dependence.

The PIM level allows for definition of domains, software components, interfaces, interaction patterns, and data elements without flattening them to a specific set of computing, communications, and middleware technologies.

Aside from enabling technology-independent design, this approach, as formalized in model-driven engineering principles, fosters interoperability.

At a minimum, a common set of interfaces and messaging standards is required for interoperability. Without a common semantic understanding of what data represent, there is significant opportunity for lack of interoperability, even if messages are correctly parsed and interfaces are followed. Therefore, a key final aspect is the recognition that data modeling is a separate, core aspect for defining interoperable systems. This aspect includes specifying definitions, taxonomies, and other semantic information to ensure there is a common understanding about what information a specific data item imparts.

This approach supports the involvement of multiple organizations in the development of one or more services, and results in increased innovation, flexibility, and improved performance. SOAs, however, constitute only one approach to implementation of OAs. Certain programs may not need SOAs. The program manager will determine the correct architecture to implement. Regardless of whether SOA approach is used, DoD has mandated an OA approach in software development. The program manager will be responsible for implementing an environment that will support OA in both programmatic and technical areas.

4.5.2. Service Repositories

DoD recognizes that there is rarely a one-size-fits-all solution to the challenging problem of software and service reuse. However, service repositories fill a growing need within DoD for commonality, reuse, and reduced duplication of effort, all of which aid interoperability by leveraging common interfaces. Programs will have access to the service repositories for their use in planning and implementation. In addition, programs will be encouraged to contribute services (within Government Data Rights constraints) to the repositories for future reuse. Where programs have requirements that cannot be met by software within the repositories, existing services may be extended to add functionality, and this approach should result in cost savings over the creation of brand new capabilities.

Constructing such repositories requires a commitment to OA, along with the adoption of existing and upcoming standards (e.g., SAE JAUS, STANAG 4586, UCS), so that a common framework exists with which to develop services. In addition, tools are necessary to ease adoption, reduce learning curves, and provide validation and certification capabilities. See Figure 8.

OA allows components to be developed once rather than redeveloped for each warfare area or mission. OA exploits software reuse and open interfaces to ensure unnecessary costs are not incurred in the redevelopment of core software. The OA approach described in this section utilizes collaboration among Government, industry, and academia to comply with principles of modularity, reusability, interoperability, affordability, and competition to develop reusable products.

Through implementation of the OA approach, DoD will develop and establish a domain service portfolio management (DSPM) repository for new acquisition and in-service programs. This repository will contain program-related software services information including standard architectures, design guidelines, service interfaces, and specifications for designing new systems or modifying existing systems. Programs will be required to consult with the DSPM repository for software reuse, where applicable. As programs design new and unique software and services, they are required to populate the DSPM repository with new information and make the service available for reuse, within Government Data Rights constraints.

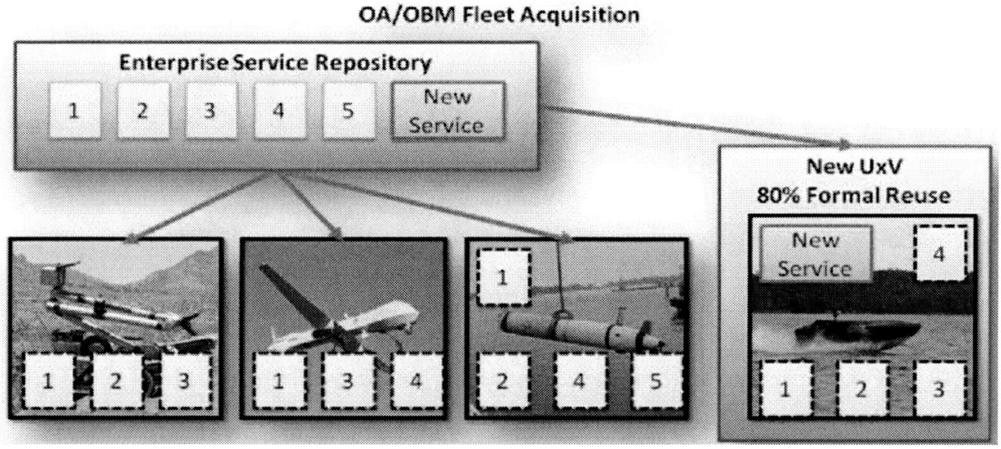

Figure 8. Cross-Domain Service Reuse Through an Enterprise Service Repository.

4.5.3. Collaborating Communities

DoD has long recognized the value in fostering collaboration between Government, industry, and academia in open forums to address interoperability and common standards. To that end, a number of integrated product teams, working groups, and other communities have formed, under the auspices and support of OUSD(AT&L), to address the interoperability challenge. These forums have enabled the Government to engage with industry at all levels, from grassroots to executive, and have enabled DoD personnel to aid in the systems and architecture design process, rather than simply being customers. These collaborating communities exist within a variety of national and international standards bodies, span the domains of unmanned systems (i.e., UAS, UGS, UMS), and address key cross-domain areas as well as domain-unique capabilities. DoD intends to continue to support this type of collaboration as it fosters the development of OAs. Current examples of these communities include the following:

- The OUSD(AT&L)-chartered UAS Task Force to coordinate critical DoD UAS issues and develop a path to enhance operations, enable interdependencies, facilitate interoperability, and streamline UAS acquisition. Within the UAS Task Force, the I-IPT has been chartered to promote UAS interoperability across the Services. The I-IPT establishes a central coordination forum for the Services' acquisition organizations and participating industry partners to share ideas that will allow the Department to build interoperability within the deployed UAS infrastructure, individual systems, and interfaces with appropriate manned weapons systems and C2 capabilities.
- Under the guidance of the Defense Science and Technology Advisory Group, the Autonomy Systems Community of Interest (CoI) closely examine the DoD's S&T investments in the enabling of autonomous systems. Specifically, this CoI identifies potential investments to advance and initiate critical enabling technology

developments and strategically assesses the challenges, gaps, and opportunities to develop and advance autonomous systems.

- The National Geospatial-Intelligence Agency (NGA) is a member of the U.S. Intelligence Community and a Department of Defense (DoD) Combat Support Agency. NGA provides support to civilian and military leaders and contributes to the state of readiness of U.S. military forces by providing geospatial intelligence (GEOINT) imagery, imagery intelligence and geospatial data (e.g., mapping, charting and geodesy), and information to ensure the knowledge foundation for planning, decision and action. NGA also contributes to humanitarian efforts, such as tracking floods and disaster support, and to peacekeeping. NGA provides unmanned systems, topographical and terrestrial data, geodesy and geophysical data, imagery and precise position and target data for unmanned system mission planning and UAS flight operations. GEOINT support includes aeronautical and safety of navigation data, vertical obstruction, digital terrain elevation data and hydrographic data.
- NATO's Joint Capability Group on Unmanned Aerial Vehicles (JCGUAV) directs interoperability efforts in unmanned aviation. JCGUAV subsumed NATO's three Military Department UAS-related groups (PG-35, Air Group 7, and Task Group 2) in 2006. Its major accomplishments to date include STANAG 4586 for UAS message formats and data protocols, STANAG 4671 for UAV Airworthiness Standard, and STANAG 7085 for the CDL communication system, which has been mandated by OSD since 1991.
- NATO's Joint Capability Group Intelligence Surveillance and Reconnaissance (JCGISR) provides interoperability between NATO and Coalition ISR systems and includes standards related to imagery formats and interfaces, data storage interfaces, motion imagery, electronic intelligence reporting, and imaging systems data links.
- Current UAS System Interoperability Profiles (USIPs), produced by the I-IPT, define the standard interface for payload products and the data link between a control station and air vehicle for line of sight (LOS) and beyond line of sight (BLOS) scenarios. Future USIPs will address other aspects of interoperability to include data encryption, additional data link technologies such as bandwidth efficient common data link (BE-CDL), and enhanced capabilities provided by future sensors.
- The Joint Architecture for Unmanned Systems (JAUS) began in 1995 as an effort by the Army's program office for UGVs in the Aviation and Missile Research, Development and Engineering Center (AMRDEC) at Redstone Arsenal to establish a common set of message formats and data protocols for UGVs. Deciding to convert JAUS to an international industry standard, the program office approached the Society of Automotives Engineers (SAE), a standards development organization (SDO) with robotics experience, which established the AS-4 Unmanned Systems Committee in August 2004. AS-4 has three subcommittees focused on requirements, capabilities, and interfaces and an experimental task group to test its recommended formats and protocols before formally implementing them. The migration to the SAE is complete, and the first set of SAE JAUS standards, focusing on the JAUS Service Interface Definition Language (JSIDL), core services, mobility services, manipulation services, and environmental sensing services, has been balloted and released. Although AS-4 is open to its members creating standards on other aspects of unmanned systems beyond message formats and data protocols for UGVs, much

of this broader work is now being undertaken by other UAS-related SDOs. STANAG 4586 is unmanned aviation's counterpart to JAUS.

- The Navy's Program Executive Officer of Littoral and Mine Warfare (PEO(LMW)) formally adopted JAUS message formats and data protocols for use with its UUVs, USVs, and UGVs in 2005. Working through SAE AS-4, the Naval Undersea Warfare Center (NUWC) expanded JAUS to serve the UMS community. It found only 21% of UMS message formats to be directly compatible with the formats of JAUS, with the high percentage of new formats needed possibly due to the operation of UMS in three dimensions versus the two dimensions of UGVs, for which JAUS was developed. UUV variants of JAUS services are in active development and have been presented to the SAE AS-4 committees.
- Under direction from the OUSD(AT&L),[11] the UAS Task Force chartered the UAS Control Segment (UCS) Working Group, which is tasked to develop and demonstrate a common, open, and scalable UAS architecture supporting UAS Groups 2 to 5 (see Fig 1. DoD UAS for Groupings). The UCS Working Group comprises Government and industry representatives and operates using a technical society model where all participants are encouraged to contribute in any area of interest. This effort incorporates the best practices of current Army, Air Force, and Navy development efforts to include, but not limited to, the following:
    - Definition of a common functional architecture, interface standards, and business rules
    - Use of open-source and Government-owned software as appropriate
    - Competitive acquisition options
    - Refinement of message sets to support all operational requirements of the systems previously defined

In addition to the definition of common capability descriptions, standards, data models, and architectures, DoD continues to promote the development of OA tools and to aid system acquisition and development in embracing the OA concepts. These efforts extend across the technology and unmanned vehicle spectrum, from software development kits, to complete architectures, addressing UGVs, UMVs, and UAVs, across all Services. Examples of such tools include:

1. The JAUS Tool Set (JTS) is a tool to help developers build JAUS-compliant software components without having to be intimately familiar with the details of JAUS. JTS allows an unmanned system designer to focus on behavior rather than messaging, protocol, and other considerations by providing a graphical user interface (GUI) service editor, validator, internal repository, C++ code generation, and hypertext markup language (HTML) document generation.
   The Navy and OSD have supported and promoted the use of the JTS and have had success incorporating it into development and acquisition efforts. Use of JTS on programs accrues benefits to a number of stakeholders in the acquisition chain and RDT&E community. These benefits include enabling a fair basis for competition among vendors so that true capabilities are evaluated; reducing vendor lock-in on unmanned systems; and enabling the development of a service repository for JAUS capabilities that have been developed and are available for reuse. JTS reduces the

threshold for entry into developing JAUS-compliant systems, opens the market to small businesses, and drives competition and innovation focused on core technology. In addition, JTS provides an accepted, common validation capability, which is critical to ensure systems maintain compliance with JAUS.

2. The STANAG 4586 Compliance Toolkit (4586CT) is an integrated set of software tools that provides passive, interactive, and automated test capability. Its core function is to verify the structure and content of data link interface (DLI) messages against both STANAG 4586 and "private" messages as defined to support service-, mission-, or platform-specific requirements. This nonintrusive capability is provided either in real time or during post-run analysis. Additionally, 4586CT can be interoperable with other DLI-compatible systems in either manual mode (where an engineer monitors and injects DLI messages into the network) or automated mode (in which 4586CT interacts directly with other DLI systems according to user-defined scripts and procedures).

   These capabilities enable 4586CT to perform compliance testing at both the message level and the higher protocol session levels of unmanned systems relative to the STANAG 4586, and other more specific interoperability profiles. Complex DLI message dialogs can be monitored and system interaction sequencing verified as 4586CT follows user-defined test programs. Because 4586CT can function as a proxy for other unmanned system components, it is also used during system development and task-specific integration testing to provide insight into unmanned system interaction and performance. Multiple instances of 4586CT can also be utilized to perform rapid prototyping of interoperation protocols during profile design; as a result, 4586CT can be a useful tool during the development of interoperability standards themselves.

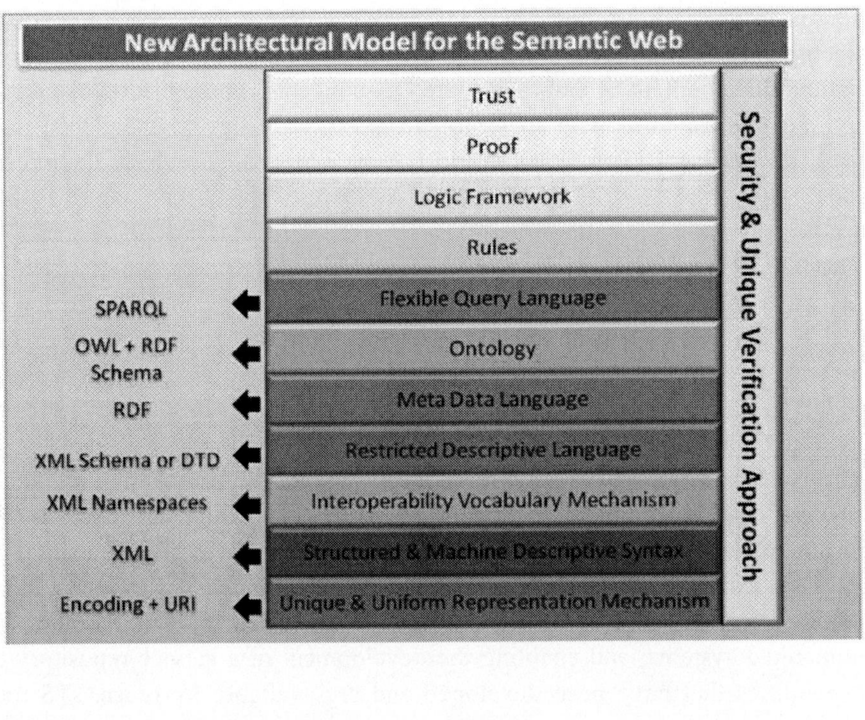

The T&E of interoperability continues to evolve with the growth of unmanned systems. The C2 of these systems presents unique test challenges as autonomous functionality expands to operating complex equipment over wireless links. The spectrum of test includes assessment of standards compliance, electromagnetic frequency testing, sensor standards, payload standards, systems interoperability, quantifiable task assessment, performance measurement, metrics development, congestion management, and performance baseline measures. The scope of operation includes operator, platform, communications grid, C2 teams, ground stations, sensor teams, and collaborating systems. While today's test represents a migration of the test challenge represented by proprietary data exchange and data formats towards service-oriented architecture exchange, it is also conceivable that the cognitive nature of unmanned systems algorithm development may actually accelerate the need for semantic knowledge exchange T&E.

The rapid acquisition of quickly evolving unmanned systems will require an unmanned systems T&E capability that evolves at a pace that exceeds this evolution. The T&E of unmanned systems interoperability requires investment in information architectures, methodology, test scenario synthesis, model-based T&E, and cross-UAS usage case repositories.

4.5.4. System Migration

Services working with OUSD(AT&L) are exploring the business case for adopting an OA approach for current and in-development systems. While various challenges still remain in the adoption of OA, the goal in the next 12 months is to identify migration pathways for all current programs of record (see Figure 9).

Quantifiable progress toward migration has already been achieved. The USAF Advanced Cockpit Block 50 (under development by General Atomics) has adopted UCS services and a common data bus. The July 2010 Block 50 implementation of takeoff, flight, payload C2, and landing utilized UCS-based software in a simulated flight environment demonstrated the architecture's utility. Additionally, Northrop Grumman has now agreed to tie its product lines into a common, open product line, with a joint mission-planning-mission-control system document signed by Northrop Grumman executives for synergy and collaboration between Broad Area Maritime Surveillance (BAMS) and Global Hawk unmanned systems. Advanced Explosive Ordnance Robotic System (AEODRS), utilizing an OA approach for hardware and software, is adopting SAE JAUS for messaging and is also developing interoperability profiles to ensure common system functionality descriptions, architectures, and data models. AEODRS is developing three classes of vehicles (dismounted operations, tactical operations, and base/infrastructure operations) with a common architecture and capability modules across the FoS. The AEODRS architecture defines the logical, mechanical, and electrical interfaces for the FoS. The AEODRS is entering Milestone B, and the development of increment 1 (dismounted operations) is now starting.

In addressing interoperability for ground systems, Robotic Systems Joint Project Office (RS-JPO) is utilizing SAE JAUS for messaging (with custom extensions as necessary) and primarily focusing on communications, payloads, power, architecture, and controller. Progress has already been made, with a modeling and simulation demonstration and an Input/Output (IO) Specification Build V1 planned for 2012.

4.6. Summary

We can no longer afford to acquire independent, proprietary unmanned systems that do not leverage interoperability. The lines in the battlespace are blurring, and the need to share information, sensors, payloads, and platforms is real. The fiscal battlespace is also blurring, and vendors must shift strategies to adhere to standards, drive toward OAs, reuse software, and develop robust repositories. The goal is to provide more capable unmanned systems to the warfighter on time, and interoperability will ultimately play a large role in this effort by enabling the composition of novel systems capabilities on a faster timescale. Figure 10 depicts an interoperability path for the future as industry and DoD strive to become more efficient.

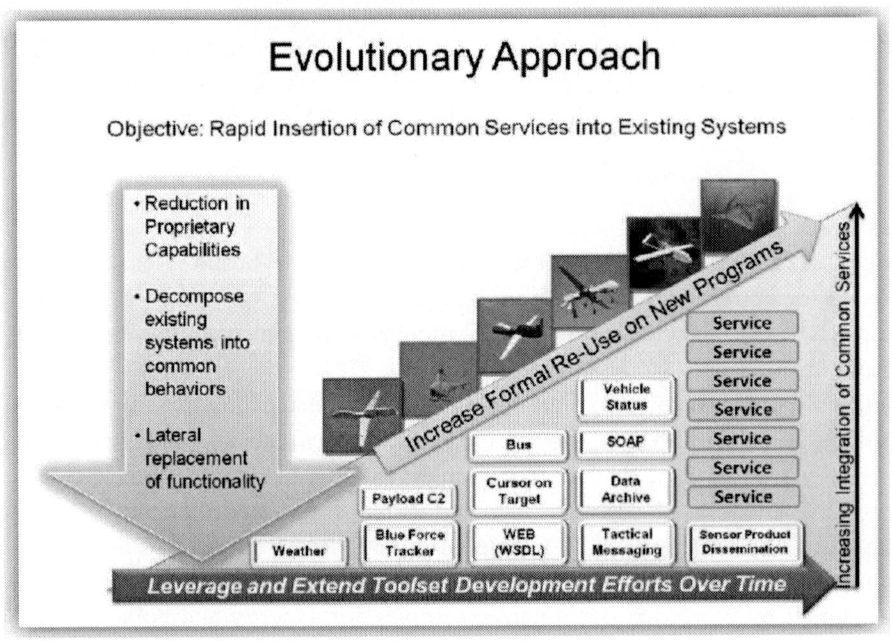

Figure 9. OA Migration Approach.

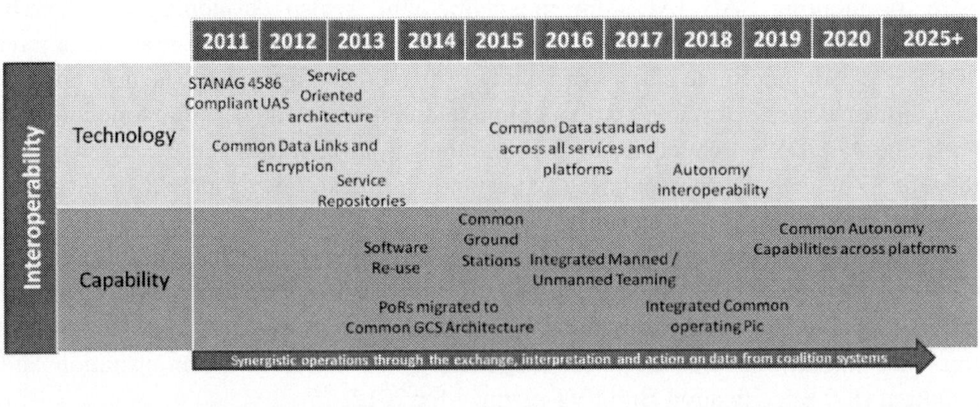

Figure 10. Interoperability Roadmap.

5. AUTONOMY

Dramatic progress in supporting technologies suggests that unprecedented levels of autonomy can be introduced into current and future unmanned systems. This advancement could presage dramatic changes in military capability and force composition comparable to the introduction of "net-centricity." DoD must understand and prepare to take maximum practical advantage of advances in this area.[12]

5.1. Functional Description

Automatic systems are fully preprogrammed and act repeatedly and independently of external influence or control. An automatic system can be described as self-steering or self-regulating and is able to follow an externally given path while compensating for small deviations caused by external disturbances. However, the automatic system is not able to define the path according to some given goal or to choose the goal dictating its path.

By contrast, autonomous systems are self-directed toward a goal in that they do not require outside control, but rather are governed by laws and strategies that direct their behavior. Initially, these control algorithms are created and tested by teams of human operators and software developers. However, if machine learning is utilized, autonomous systems can develop modified strategies for themselves by which they select their behavior. An autonomous system is self-directed by choosing the behavior it follows to reach a human-directed goal. Various levels of autonomy in any system guide how much and how often humans need to interact or intervene with the autonomous system, and these levels will be discussed shortly. In addition, autonomous systems may even optimize behavior in a goal-directed manner in unforeseen situations (i.e., in a given situation, the autonomous system finds the optimal solution).

The special feature of an autonomous system is its ability to be goal-directed in unpredictable situations. This ability is a significant improvement in capability compared to the capabilities of automatic systems. An autonomous system is able to make a decision based on a set of rules and/or limitations. It is able to determine what information is important in making a decision. It is capable of a higher level of performance compared to the performance of a system operating in a predetermined manner.[13]

5.2. Today's State

In 2010, the USAF released the results of a year-long study highlighting the need for increased autonomy in modern weapon systems, especially given the rapid introduction of UAS. This study, "Technology Horizons," identified the need for greater system autonomy as the "*single greatest theme*" for future USAF S&T investments. The study cited the potential for increased autonomy to improve effectiveness through reduced decision cycle time while also enabling manpower efficiencies and cost reductions.

Autonomous capabilities have been enabled by advances in computer science (digital and analog), artificial intelligence, cognitive and behavioral sciences, machine training and learning, and communication technologies. In order to achieve operational acceptance and trust of these autonomous capabilities in the highly dynamic unmanned system environment, improvement is essential in advanced algorithms that provide robust decision-making capabilities (such as machine reasoning and intelligence); automated integration of highly disparate information; and the computational construct to handle data sets with imprecision, incompleteness, contradiction, and uncertainty.

In response to CCDR needs, the USAF has aggressively expanded UAS capabilities to a target of 65 combat air patrols (CAPs). According to the USAF, 1750 pilots from the Total Force (Active, Guard, and Reserve) are required to maintain these CAPs, which operate around the clock. This increasing manpower requirement is occurring at a time when constrained budgets are limiting growth in Service manpower authorizations. This challenge is not limited to the USAF, but is facing all the military Services. Today's unmanned systems require significant human interaction to operate. As these systems continue to demonstrate their military utility, exploit greater quantities of intelligence, and are fielded in greater numbers, the demand for manpower will continue to grow. The appropriate application of autonomy is a key element in reducing this burden.

> Our Program Managers should be scrutinizing every element of program cost, assessing whether each element can be reduced relative to the year before, challenging learning curves, dissecting overheads and indirect costs, and targeting cost reduction with profit incentive—in short, executing to what the program should cost.
>
> *–Under Secretary of Defense Memorandum for Acquisition Professionals, Better Buying Power, September 2010*

5.3. Problem Statement

The increased manpower to operate unmanned systems is adding stress to the overall workload of the armed forces. This stress highlights the need to transition to a more autonomous, modern system of warfare. The USAF Chief of Staff General Norton Schwartz emphasized the need for more automation in the following statement:[14]

> [The trend] cannot continue indefinitely. There is a place for automation here that reduces the manpower requirement, both to operate and to process the backend data stream.
>
> *– Gen Norton Schwartz, Air Force Chief of Staff*

For unmanned systems to fully realize their potential, they must be able to achieve a highly autonomous state of behavior and be able to interact with their surroundings. This advancement will require an ability to understand and adapt to their environment, and an ability to collaborate with other autonomous systems, along with the development of new verification and validation (V&V) techniques to prove the new technology does what it should. Each of these topics is discussed in more detail below. Advances in autonomy at the system level must proceed with awareness of potential disadvantages and vigilance for unintended consequences, which may include diminished command over parts of the forces structure. Every operation includes rules of engagement, air tasking order (ATO)/special instructions (SPINS), and options of dynamic changes of command direction; and intent must not be traded off. The ability to respond to the unexpected cannot be diminished. For example, dealing with volcanic ash in the atmosphere cannot be reliably predicted at the beginning of an eruption. The ability to respond and avoid affected airspace is an example of a condition that may be difficult for autonomy to address. Implementing autonomy can lead to a loss of human attention to vital oversight in matters having potentially dangerous or lethal consequences. Caution must be used at the system-of-systems level, and constraints applied in some operations in order to allow autonomy in others. Finally, surrendering decision trust to a software-based and self-learning design outside the context of specific operations is a matter of high rigor, and must be examined in the context of organizational-unit and theater CONOPS.

5.4. Way Ahead

Significant advances have been made in autonomy, but many challenges still exist. For relatively static environments and undemanding missions and objectives, rule-based autonomous systems can be highly effective. However, most DoD environments and mission tasks dictate that unmanned systems operate in complex and uncertain environments as well as possess the ability to interact and collaborate with human operators and human teammates. Additionally, autonomous systems need the capability to interact and work together with other autonomous systems, to adapt to and learn from changes in the environment and missions, and to do so safely and reliably. One goal of automation is to leap forward in capabilities using human augmentation. Automated assistance of whatever kind does not simply enhance our ability to perform the task: it changes the nature of the task itself.[15]

Table 3. Four Levels of Autonomy

Level	Name	Description
1	Human Operated	A human operator makes all decisions. The system has no autonomous control of its environment although it may have information-only responses to sensed data.
2	Human Delegated	The vehicle can perform many functions independently of human control when delegated to do so. This level encompasses automatic controls, engine controls, and other low-level automation that must be activated or deactivated by human input and must act in mutual exclusion of human operation.
3	Human Supervised	The system can perform a wide variety of activities when given top-level permissions or direction by a human. Both the human and the system can initiate behaviors based on sensed data, but the system can do so only if within the scope of its currently directed tasks.
4	Fully Autonomous	The system receives goals from humans and translates them into tasks to be performed without human interaction. A human could still enter the loop in an emergency or change the goals, although in practice there may be significant time delays before human intervention occurs.

5.4.1. Transcending to Higher Levels of Autonomy

Autonomy reduces the human workload required to operate systems, enables the optimization of the human role in the system, and allows human decision making to focus on points where it is most needed. These benefits can further result in manpower efficiencies and cost savings as well as greater speed in decision making. Autonomy can also enable operations beyond the reach of external control or where such control is extremely limited (such as in caves, under water, or in areas with enemy jamming or degraded communications). Advances in autonomy will further increase operational capability, manpower efficiencies, and cost savings.

> … the ability to understand and control future costs from a program's inception is critical to achieving affordability requirements.
>
> *–Under Secretary of Defense Memorandum for Acquisition Professionals, Better Buying Power, September 2010*

While reduced reliance on human operators and analysts is the goal of autonomy, one of the major challenges is how to maintain and facilitate interactions with the operator and other human agents. An alternative statement of the goal of autonomy is to allow the human operator to "work the mission" rather than "work the system." In other words, autonomy must be developed to support natural modes of interaction with the operator. These decision-making systems must be cognitively compatible with humans in order to share information states and to allow the operator and the autonomous system to interact efficiently and effectively. The level of autonomy should dynamically adjust based on workload and the perceived intent of the operator. Common terms used for this concept are *sliding autonomy* or *flexible autonomy*. The goal is not about designing a better interface, but rather about designing the entire autonomous system to support the role of the warfighter and ensure trust in the autonomy algorithms and the system itself. Table 3 contains the most commonly referenced description of the levels of autonomy that takes into account the interaction between human control and the machine motions.

> The single greatest theme to emerge from "Technology Horizons" is the need, opportunity, and potential to dramatically advance technologies that can allow the Air Force to gain the capability increases, manpower efficiencies, and cost reductions available through far greater use of autonomous systems in essentially all aspects of Air Force operations. Increased use of autonomy — not only in the number of systems and processes to which autonomous control and reasoning can be applied but especially in the degree of autonomy that is reflected in these — can provide the Air Force with potentially enormous increases in its capabilities, and if implemented correctly can do so in ways that enable manpower efficiencies and cost reductions.
>
> *– USAF Report on Technology Horizons: A Vision for Air Force Science and Technology During 2010-2030, 15 May 2010*

5.4.2. Ability to Understand and Adapt to the Environment

To operate in complex and uncertain environments, the autonomous system must be able to sense and understand the environment. This capability implies that the autonomous system must be able to create a model of its surrounding world by conducting multisensor data fusion (MDF) and converting these data into meaningful information that supports a variety of decision-making processes. The perception system must be able to perceive and infer the state of the environment from limited information and be able to assess the intent of other agents in the environment. This understanding is needed to provide future autonomous systems with the flexibility and adaptability for planning and executing missions in a complex, dynamic world.

Although such capabilities are not currently available, recent advancements in computational intelligence (especially neuro-fuzzy systems), neuroscience, and cognition science may lead to the implementation of some of the most critical functionalities of heterogeneous, sensor net-based MDF systems. The following developments will help advance these types of processing capabilities:

1. *Reconfigurability of sensor weighting:* When a heterogeneous sensor net is used for an MDF system, each sensor has a different weight for different applications. As an example, regardless of whether a dissimilar MDF methodology is used to identify an object, an image sensor has much higher weight than radar. On the other hand, when an MDF methodology is used to measure a distance from the sensor to an object, a rangefinder or radar has a much higher weight than an image sensor.
2. *Adaptability of malfunctioning sensors and/or misleading data:* Even if an MDF methodology is used to identify an object, an image sensor cannot perform if it is faced to the sun. Data from the image sensors will either be saturated or need to be calibrated. Additionally, the image sensor data needs to be continuous calibrated if the weather is cloudy and changing because the measured data will be different based on shadows and shading. Therefore, the environment of a heterogeneous sensor net is a key parameter to be considered for design and implementation of an MDF system.
3. *Intelligent and adaptive heterogeneous data association:* Heterogeneous, sensor net-based MDF systems must process different data simultaneously, such as one-dimensional radar signals, two-dimensional imaging sensor data, etc. As the combination of heterogeneous sensors change, the data combination is changed. Therefore, adaptive data association must be performed before conducting MDF and data input to the decision-making module.
4. *Scalability and resource optimization of self-reconfigurable fusion clusters:* The limiting factor of an MDF system is the scalability of self-reconfiguring the fusion cluster to adapt to a changing battlefield and/or the malfunction of one or more sensors. As the number of sensors used for a sensor net increases, the combinatorial number of reconfigurations exponentially increases. To manage such complexity, the MDF system will require a highly intelligent, fully autonomous, and extremely versatile reconfigurable algorithm, including sensor resource management and optimization. Great progress has been made in sensor management algorithms and cross-cued sensor systems, but true optimization is an elusive goal that is currently unavailable. Such capability can be obtained only from intelligent computing technology, which is currently in its infancy.

While robustness in adaptability to environmental change is necessary, the future need is to be able to adapt and learn from the operational environment because every possible contingency cannot be programmed *a priori*. This adaptation must happen fast enough to provide benefits within the adversary's decision loop, and the autonomy should be constructed so that these lessons can be shared with other autonomous systems that have not yet encountered that situation. Yet even in a hostile, dynamic, unstructured, and uncertain environment, this learning must not adversely affect safety, reliability, or the ability to collaborate with the operator or other autonomous systems. The flexibility required of autonomous systems in dynamic, unstructured environments complicates the predictability needed for U.S. commanders to "trust" the autonomy.

"Trust" will be established through robust operational T&E along with safeties and safeguards to ensure appropriate behavior. Complex autonomous systems must be subject to rigorous "red team" analysis in order to evaluate the full range of behaviors that might emerge in environments that simulate real-world conditions. Safeties and safeguards are also required to mitigate the consequences of failures. Because artificial systems lack the human ability to step outside a problem and independently reevaluate a novel situation based on commander's intent, algorithms that are extremely proficient at finding optimal solutions for specific problems may fail, and fail badly, when faced with situations other than the ones for which they were programmed. Robust safeties and control measures will be required for commanders to trust that autonomous systems will not behave in a manner other than what is intended on the battlefield.

5.4.3. Enabling Greater Autonomy in TPED Processes

In addition to C2 processes, traditional TPED processes offer huge opportunities for reducing the degree of human involvement. Near-term developments could introduce a greater degree of automation, ultimately evolving to more autonomous systems. Current TPED processes are manpower intensive. In today's combat environment, most full-motion video (FMV) and still imagery is monitored and used in real time, but then stored without being fully analyzed to exploit all information about the enemy. This challenge is not unique to the unmanned environment, but it has been exacerbated by the large numbers of ISR-capable, long-endurance unmanned systems being fielded. These systems are collecting great quantities of information and overwhelming current TPED processes. Near-term steps might include implementation of change detection and automatic target recognition software to enable automated cueing that identifies and calls attention to potential threats. Applications of face recognition software could enable high-fidelity FMV to identify individuals of interest. Increased automation in communications intelligence sensors has the potential to identify key words and even specific voices to rapidly alert operators to targets of interest. Ultimately, automated cross-cueing of different sensor types in a networked environment could enable greater autonomy in tasking systems and their sensors to identify and track threats more rapidly.

Increased processing power and information storage capacities also have the potential to change how unmanned systems operate. For example, many current UAS transmit ISR data that is processed and exploited in ground stations. If more processing and exploitation processes can be accomplished onboard a UAS (like the automatic target recognition or communications intelligence examples discussed above), the system can disseminate actionable intelligence for immediate use and reduce bandwidth requirements. FMV ISR, for example, uses roughly an order of magnitude more bandwidth than the C2 data for a UA. By accomplishing more of the TPED process onboard the unmanned system, the link bandwidth can then be focused on transmitting only what's needed, and the overall bandwidth requirements can be reduced.

> Today an analyst sits there and stares at Death TV for hours on end trying to find the single target or see something move or see something do something that makes it a valid target. It is just a waste of manpower. It is inefficient!
>
> – *Gen James Cartwright, Vice Chairman of the Joint Chiefs of Staff, during remarks to the U.S. Geospatial Intelligence Foundation on 4 Nov 2010*

5.4.4. Ability to Collaborate with Other Autonomous Systems

In addition to understanding the environment, unmanned systems must also possess the ability to collaborate through the sharing of information and deconfliction of tasking. Collaborative autonomy is an extension of autonomy that enables a team of unmanned systems to coordinate their activities to achieve common goals without human oversight. This trend in autonomy will continue to reduce the human role in the system. Autonomously coordinated unmanned systems may be capable of faster, more synchronized fire and maneuver than would be possible with remotely controlled assets. This trend will lead to a shift toward strategic decision making for a team of vehicles and away from direct control of any single vehicle.

The ability to collaborate is one of the keys to reducing force structure requirements. The collaborative autonomy that is developed must be scalable to both larger numbers of heterogeneous systems as well as increased mission and environment complexity. Collaborative autonomy must be able to adapt to the air, ground, and maritime traffic environment and to changes in team members, operators, and the operational environment.

5.4.5. Development of New Approaches to Verification and Validation (V&V)

To ensure the safety and reliability of autonomous systems and to fully realize the benefits of these systems, new approaches to V&V are required. V&V is the process of checking that a product, service, or system meets specifications and that it fulfills its intended purpose. These components are critical in a quality management system such as ISO 9000. Today's V&V processes will be severely stressed due to the growth in the amount and complexity of software to be evaluated. They utilize existing industry standards for software certification that are in place for manned systems (e.g., DO-178B). Without new V&V processes, such as the use of trust audit trails for autonomy, the result will be either extreme cost growth or limitations on fielded capabilities.

Efforts leading to advancements in computational intelligence as well as the appropriate V&V processes are essential. Enhanced V&V technologies would provide both near-term cost reduction and enhanced capabilities for current autonomous systems and would enable otherwise cost-prohibitive capabilities in the future. New autonomous system test and analysis capabilities are also required to assess intelligent single-vehicle and group behaviors.

These technological enhancements and policy actions would lead to more effective development, testing, and operations of current and future autonomous systems.

5.4.6. Policy Guidelines to Ensure Safe Operation

Additional measures, beyond V&V, will be required to ensure safe operation of autonomous systems. No V&V process can guarantee 100% error-free operation of complex systems. As software complexity increases, predicting the precise behavior of autonomous systems in real-world environments will be increasingly difficult. Policy guidelines are necessary in order to ensure that if failures or malfunctions occur, or if an unmanned system encounters an unanticipated situation, the system continues to operate appropriately.

Policy guidelines will especially be necessary for autonomous systems that involve the application of force. Current armed, unmanned systems deploy lethal force only in a fully human-operated context (level 1) for engagement decisions. For these systems, the decisions both to employ force and to choose which specific target to engage are made by a human. The United States does operate defensive systems for manned ships and installations that have human-supervised autonomous modes (level 3), and has operated these systems for decades. For the foreseeable future, decisions over the use of force and the choice of which individual targets to engage with lethal force will be retained under human control in unmanned systems.

5.5. Summary

Technological advances in autonomy are critical as the need to field greater numbers of unmanned systems stresses the limited number of available operators. Challenges in the area of autonomy address not only functionality, but also transparency to the operator, safety, and reliability. Figure 11 provides a vision into the future of the autonomy advances that are required to maintain an affordable force structure and confidently operate unmanned systems in an increasingly complex environment. Initially, autonomy will improve the safe operations of unmanned systems within the increasingly complex environment of military operations as well as reduce operator workload associated with mundane and noncritical processes. Ultimately, autonomy will increase warfighter effectiveness by enhancing unmanned systems capabilities and expanding their capacity to effect results in the battlespace.

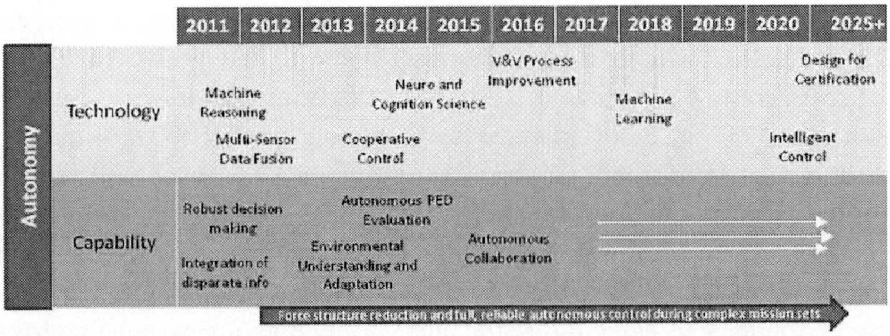

Figure 11. Autonomy Roadmap.

6. AIRSPACE INTEGRATION (AI)

6.1. Functional Description

Over the past several years, UAS have become a transformational force multiplier for DoD. The numbers and roles of UAS have expanded dramatically to meet mission demands, and operational commanders have come to rely upon robust and persistent ISR support from unmanned platforms executing their core missions against hostile forces. DoD UAS require routine NAS access in order to execute operational, training, and support missions and to support broader military and civil demands. UA will not achieve their full potential military utility to do what manned aircraft do unless they can go where manned aircraft go with the same freedom of navigation, responsiveness, and flexibility. Military aviation is a major contributor to the virtue of maneuver for our forces in warfare.

While the force structure continues to grow, the ability to integrate UAS into the NAS has not kept pace. Current access for UAS is greatly limited primarily due to FAA regulatory compliance issues that govern UAS operations in the NAS. DoD UAS operations conducted outside of restricted, warning, and prohibited areas are authorized only under a (temporary) COA from the FAA. Similar issues need to be resolved for access to international and foreign national airspace.

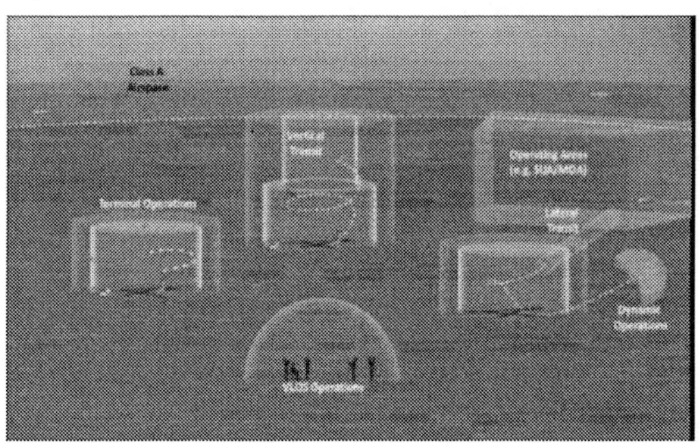

The *DoD UAS Airspace Integration Plan, March, 2011* provides a more comprehensive discussion on the topic of AI. In this plan, DoD provides an incremental approach strategy to provide DoD UAS access to a given operations profile that leads to a full dynamic operations solution. This methodology recognizes that DoD requires access to differing classes and types of airspace as soon as possible and that routine dynamic operations will likely take several years to implement. Figure 12 depicts the six access profiles.

6.1.1. Vision

DoD's vision is to ensure UAS have routine access to the appropriate airspace required to meet mission needs. For military operations, UAS will operate with manned aircraft using CONOPS that make manned or unmanned aircraft distinctions transparent to air traffic services (ATS) authorities and airspace regulators. Having robust UAS AI capabilities for all classes of U.S. airspace is fundamental to flexible worldwide UAS deployment.

NAS ACCESS REQUIREMENTS

- Aircraft must be Airworhty
- Must be operated by a Qualified Pilot/ Operator
- Compliant with Operating Rules, Standards, and Procedures

> … it is vital for the Department of Defense and the Federal Aviation Administration to collaborate closely to achieve progress in gaining access for unmanned aerial systems to the National Airspace System to support military requirements.
>
> – *110th Congress, NDAA for FY09, Sect 1039*

6.1.2. Precepts

The *2010 DoD Airspace Integration Plan* outlines DoD's approach, which is summarized by four overarching precepts (see right). The U.S. military will use its vast experience to develop the safest, most capable UAS fleet possible. We will strive for maximum compliance with existing regulatory guidance and inform regulatory processes when changes are needed. DoD will fully leverage statutory authorities to design, test, and ultimately certify its UAS in compliance with applicable standards, regulations, and orders. The regulatory and policy changes may be broad in scope to affect multiple Military Departments and Combatant Commands (CCDRs); therefore, UAS AI activities should make every effort to be coordinated prior to engaging with FAA or other external agencies.

AIRSPACE INTEGRATION PERCEPTS

- Applu Our World-leading Aviation Expertise to UAS
- Conform Where Possible, Create Where Needed
- Leverage DOD Authorities and Equities
- Engage as One

6.2. Today's State

In order for any military aircraft — manned or unmanned — to fly routinely in domestic and international airspace, three foundational requirements must be met. These three requirements are essential and form the foundation for UAS AI. Title 10 of the United States Code (USC) is the legal underpinning for the roles, missions, and organization of DoD and provides authority for the military departments to organize, train, and equip U.S. forces to fulfill the core duties for national defense. Consistent with this statutory authority and longstanding practice and reinforced by interagency agreements, DoD is responsible for establishing airworthiness and pilot training/qualification requirements for the military and ensuring rigorous military standards are satisfied.[16] The third and most complex requirement, regulatory compliance, encompasses both internal military department regulations and external FAA and International Civil Aviation Organization (ICAO) flight regulations.

DoD Aircraft and Pilots/Operators are Certified by DOD

6.2.1. Airworthiness

Airworthiness is a basic requirement for any aircraft system, manned or unmanned, to enter the NAS. The primary guidance for DoD airworthiness certification is found in MIL-HDBK-516B, *Airworthiness Certification Criteria.* This document defines airworthiness as "*the ability of an aircraft system/vehicle to safely attain, sustain and terminate flight in accordance with an approved usage and limitation.*"[17] Airworthiness certification ensures that DoD aircraft systems are designed, manufactured, and maintained to enable safe flight. Certification criteria, standards, and methods of compliance establish a minimum set of design and performance requirements for safely flying a given category and class of aircraft. The DoD is expanding current military airworthiness guidance to include criteria that address those component and system attributes that are unique to UA. UAS-unique standards derived from NATO STANAGs (e.g., 4671[18], 4705, and 4703) will be reviewed and incorporated as appropriate.

6.2.2. Pilot/Operator Qualification

The DoD determines where and how it will operate its aircraft, and each Military Department creates the qualification training programs necessary to safely accomplish the missions of that aircraft or weapon system. The standards to train and qualify pilots/operators of UAS will remain under the authority of the Military Departments and appropriate CCDRs. UAS pilot/operator training requires a different skill set from the set needed for flying manned aircraft due to differences such as the means of takeoff, cruising, and landing by visual remote, aided visual, or fully autonomous methods. Therefore, the Military Departments and CCDRs must apply the minimum training standards outlined in CJCSI 3255.01 to their respective training programs to ensure the requisite knowledge, skills, and abilities are addressed appropriately.

6.2.3. Regulatory Compliance

The Military Departments have a robust process for establishing manned aircraft flight standards and procedures. However, the current ambiguity and lack of definition in national and international regulatory guidelines and standards for UAS make it difficult to know, with consistency or certainty, whether UAS can comply. In fact, some current UAS may already be operating at appropriate levels of safety; however, until the necessary UAS-specific standards, regulations, and agreed-upon compliance methodologies are defined, establishing regulatory compliance for more routine operations is difficult. In the meantime, UAS operations within the NAS are treated as exceptions through the COA process.

While many requirements can be met through the use of existing manned aircraft, many missions are more efficiently and safely accomplished by using unmanned platforms. Technology advancements may be able to help resolve regulatory compliance issues for UA (particularly Title 14 of the U.S. Code of Federal Regulations (14 CFR) 91.113 containing the see and avoid provision); however, the level and complexity of technology required to resolve today's regulatory compliance issues will negatively affect system affordability.

6.3. Problem Statement

The number of UAS in the DoD inventory is growing rapidly. The increase in numbers, as well as the expanding roles of UAS, has created a strong demand for access to national and international airspace and has quickly exceeded the current airspace available for military operations.

6.4. Way Ahead

6.4.1. Methodology

DoD's UAS NAS access methodology includes the array of UAS platform capabilities, required airspace, technology improvement, and implementation activities/products required to attain routine operations within the NAS. This methodology uses an incremental approach to provide DoD UAS access to a given operations profile that leads to a full dynamic operations solution (see Figure 13). This methodology recognizes that DoD requires access to differing classes and types of airspace as soon as possible and that routine dynamic operations will likely take several years to implement.

The profiles, as outlined in an operational view (Figure 12) and DoD's AI Plan, may be used individually to access specific local airspace or integrated together to satisfy additional airspace requirements. Visual LOS operations establish a means to conduct UAS operations in Visual Flight Rules conditions. The terminal area profile is intended to facilitate UAS operations in a confined volume of airspace, such as Class D airspace or near restricted airspace. UAS operating areas, such as special use airspace (e.g., restricted area, or military operations area (MOA)), can be accessed either by flying through a lateral corridor (through Class E) or by vertically ascending to Class A airspace and flying across. While operating areas are limited to restricted or warning areas, MOAs are desirable because they offer a wide variety of airspace spanning 43 states to provide a robust, nationwide UAS training capability without the creation of new airspace.

Plans and programs to enable UAS operations within a profile will be evaluated for joint applicability and NAS access utility. For example, because most of the required near-term airspace for DoD UAS will be in Class D, E, and G, DoD intends to focus much of its near-term resources on addressing this major need.

6.4.2. Policy

Policy agreements can maintain the safety of the NAS while also allowing certain requirements to be fulfilled. In 2007, DoD and FAA signed a Memorandum of Agreement (MoA) allowing limited UAS operations for small UAS below 1200 ft above ground level (AGL) and UAS within DoD-controlled, non-joint-use Class D airspace. The 2007 MoA will be updated periodically, as needed, to allow DoD to incrementally increase access to the NAS. For example, the small UAS special federal aviation regulation is expected to be complete in 2013, but DoD can immediately leverage this work by seeking to incorporate many of the recommendations into an updated DoD-FAA MoA.

The Policy Board on Federal Aviation (PBFA) serves as the DoD liaison with the FAA on federal air traffic control and airspace management. The board provides policy and planning guidance to ensure the Military Departments have sufficient airspace to fulfill requirements. With support from OSD and the Military Departments, the board is working with FAA to update the 2007 MoA.

Where broader issues involve other agencies, the DoD participates in the UAS Executive Committee (ExCom). The ExCom acts as a focal point for senior leaders from FAA, DoD, DHS, and National Aeronautics and Space Administration (NASA) to meet periodically to resolve any policy and procedural disputes and to identify solutions to the integration of UAS into the NAS. The ExCom has established a working group to address COA issues, published a plan to Congress in October 2010, and continues to address specific issues such as collecting and sharing UAS safety data.

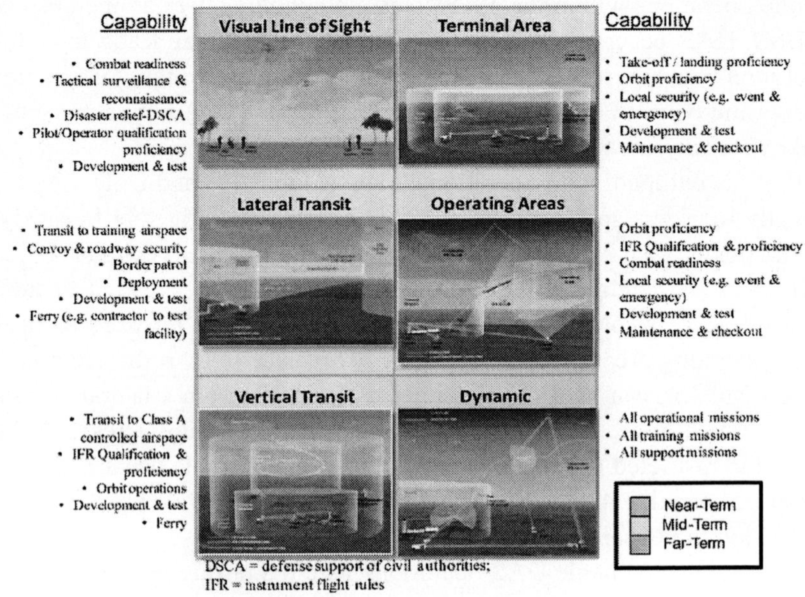

Figure 13. Incremental Approach to Regulatory Compliance.

6.4.3. Technology

Current UAS are built to different specifications for different purposes; therefore, showing individually that each system is safe for flight in the NAS can be complicated, time consuming, and costly. Routine access cannot happen until DoD and FAA agree to an acceptable level of safety for UAS, and the appropriate standards are developed to meet that threshold. With developed standards, UA will be operationally treated as manned systems, and such treatment will improve interoperability with other systems, cost savings, and development transparency.

Until those necessary UAS-specific standards are established, requirements will be dependent on the individual system and intended flight environment (access profiles). Each system's mission requirements will drive the selection of sense and avoid (SAA) solutions and process for implementation. Ground-based sense and avoid (GBSAA) can provide an initial means to maintain aircraft separation requirements for multiple profiles, while improvements to sensor and automation technology will continue to improve an airborne SAA (ABSAA) solution.

GBSAA efforts are focused on developing methods to provide aircraft separation within a prescribed volume of airspace using a ground-based system that includes sensors, displays, communications, and software. GBSAA solutions will incrementally relieve restrictions on existing COAs and facilitate UAS training and operations in the NAS. This effort is establishing requirements, gathering data, performing modeling and simulation, testing and verifying collected data, and obtaining airworthiness approvals, as appropriate. GBSAA can particularly benefit smaller UAS where other SAA solutions are cost prohibitive.

ABSAA efforts are focused on developing onboard capability to perform both self-separation and collision avoidance that ensure an appropriate level of safety. Current programs have phased validation schedules for due regard, en-route/Class A, and divert/ Class E/G operations as technology innovation and integration allow. GBSAA and ABSAA may be applied as a single or combined solution to some access profiles to maximize safety and/or reduce operational costs.

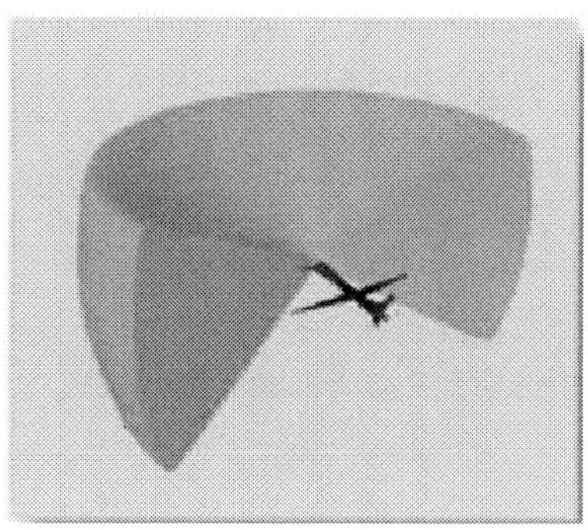

6.4.4. CONOPS Development

The DoD is developing an AI CONOPS to provide a framework for common UAS practices, procedures, and flight standards in NAS and international airspace. It is intended to standardize UAS access methodologies and procedures, implement appropriate methods for compliance with see-and-avoid requirements, and inform development of an UAS AI Initial Capabilities Document (ICD). It will establish a standard suite of lost-link, lost-communications, and lost-SAA procedures for DoD UAS in all phases of flight. These procedures will help define methods for notification and the appropriate action to either regain link or recover/divert the UA. The CONOPS also provides the operational and procedural construct to employ the access profiles at bases across the United States and to inform the process of basing UAS in locations outside the continental United States (OCONUS).

6.4.5. Requirements Development

The CONOPS, along with the Military Departments' individual location airspace requirements, will feed development of an UAS AI ICD. The UAS AI ICD is intended to identify the financial requirements for UAS integration into the NAS across the United States and OCONUS. As the initial SAA technologies mature through development and validation, they can be applied to the appropriate profiles and documented in the UAS AI ICD. This effort will allow the Military Departments to accurately estimate the costs to operate UAS at any given individual location as needed.

> … limited access to airspace is having a negative impact on the unmanned aviation community and many regions of the U.S. that are ready to support UAS industry growth.… over the next 15 years more than 23,000 UAS jobs could be created in the U.S. as the result of UAS integration into the NAS.
>
> – *Aerospace Industries Association. (2010). Total Employment: Annual Calendar Years 1990-2009. Available at AIA website: http://www.aia-aerospace.org/assets/stat12.pdf*

6.4.6. Timing of Activities

The DoD is focusing on near-term, mid-term, and far-term activities. This timing allows for immediate improvements in NAS access, while working toward viable long-term solutions.

- Near-term activities address small UAS, DoD-controlled airspace, and operations under COAs. Priority is given to initiatives that reduce COA requirements and streamline the FAA approval process. DoD believes significant near-term improvement in UAS NAS access is achievable through COA, policy, and procedural initiatives.

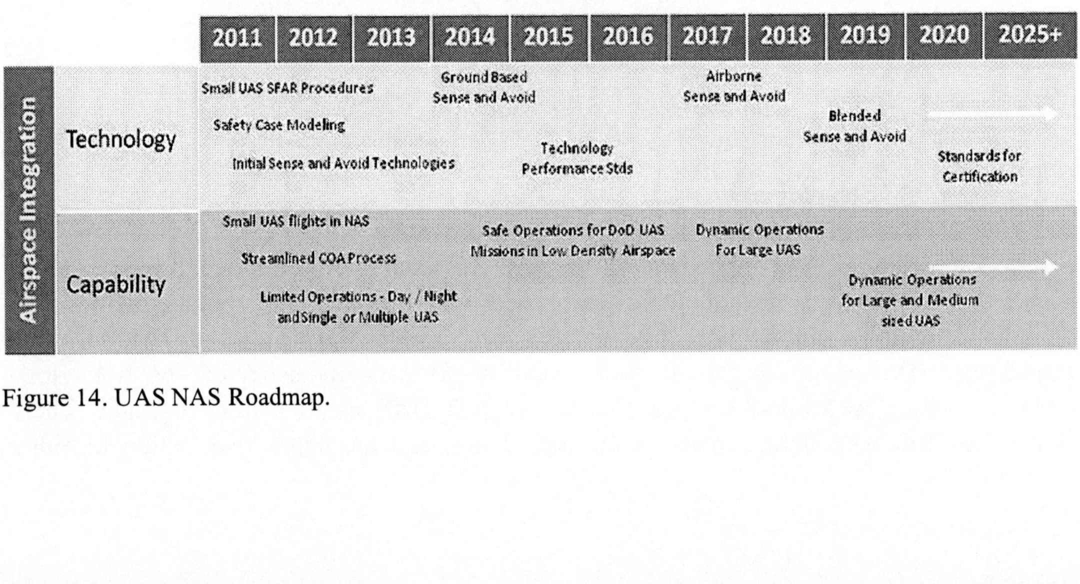

Figure 14. UAS NAS Roadmap.

- Mid-term activities address local airfield and transit operations. Where policy and procedures fall short of achieving the long-term objective of routine access, a significant investment in standards and technology development is necessary. Priority is given to developing validated AI requirements and associated standards and to establishing an SAA capability that will provide NAS access through special rules and policy, new procedures, and use of ground-based sensor technology.
- Far-term activities address most UAS missions in any operating location and airspace to include FAA's Next Generation Air Traffic Control System (NextGen). The end state is routine NAS access comparable to manned aircraft for all DoD UAS.

6.5. Summary

DoD UAS have become a critical component of military operations. Many DoD UAS now require rapidly expanded access to the NAS and international civil airspace to support operations, training, testing, and broader governmental functions.

In order for military aircraft to fly routinely in domestic and international airspace, the aircraft must be certified as airworthy, operated by a qualified pilot/operator in the appropriate class(es) of airspace, and comply with applicable regulatory guidance. DoD exercises sole certification authority for its aircraft and pilots/operators, consistent with authority provided in Title 10 of the US Code.

DoD's UAS NAS access methodology uses an incremental approach to provide DoD UAS critical access to a given operations profile prior to implementing a full dynamic operations solution. DoD's immediate focus is gaining near-term mission-critical access while simultaneously working toward far-term routine NAS access. DoD's efforts will have positive affordability effects by championing utilization of UAS within the NAS. This progress will be accomplished through policy and procedural changes as well as technology and standards development and is thoroughly outlined in the AI Plan. The end state is routine NAS access comparable to manned aircraft for all DoD UAS operational, training, and support missions.

7. COMMUNICATIONS

> "Ongoing operations in Southwest Asia continue to drive the voracious demand for pilots, support personnel and bandwidth above all"
>
> – *Col. J.R. Gear, USAF, speaking at recent C4ISR Journal Conference in Washington.*

7.1. Functional Description

DoD unmanned systems need a process for operational control and mission data distribution, especially for nonautonomous systems. For some ground and maritime systems, these types of exchanges of information can use a cable for the transmission path, but for highly mobile unmanned operations, the exchange is more likely to use signals sent across the electromagnetic spectrum (EMS) or by other means (e.g., acoustical or optical). The EMS is highly regulated at the national[19] and international[20] levels. While numerous over-the-air communication systems have been designed, built, and fielded and have performed reasonably well, others have been fielded in a noncompliant status and have not met difficult operational constraints.

DoD's desire is to operate unmanned systems in theater or within the United States and its possessions so that communication constraints do not adversely affect successful mission execution. Specifically, DoD must significantly improve communication transmission efficiencies; attain better bandwidth efficiencies; increase transmitter and receiver efficiencies; and acquire communications systems that are of less size and weight, require less power, and provide more efficient cooling to operate.

The operational employment of Unmanned Aircraft Systems requires access to a range of SATCOM capabilities. Planning and budgeting for UAS operations must take into account realistic assessments of projected SATCOM bandwidth (both military and commercial) in a range of operational scenarios. Investments in UAS systems must be matched with appropriate investments in the military and commercial SATCOM capabilities that are required to support UAS operations.

7.2. Today's State

The state of unmanned systems communication systems differs greatly among the air, ground, and maritime environments. In supporting operations in OIF, Operation New Dawn (OND), and OEF there has been a large number of new sensors and communication systems installed on various fielded unmanned systems. These have significantly increased the amount of data that has been collected, and that is desired to be sent to local and remote warfighters. To get the needed data to the remote warfighters, the DoD pays significant funds to several commercial large data transmission companies. Many current unmanned systems have experienced the impact of frequency congestion, interference from systems operating in adjacent frequency bands, and the physical limits associated with the spectrum that has been made available.

The following paragraphs describe the current communication environments by domain.

7.2.1. Unmanned Ground Systems (UGS)

Until recently, most unmanned systems utilized several radios: one for data, one for video, and sometimes one for voice. Because of congestion, frequency competition, and regulatory challenges in several theaters, many of these communication systems were redesigned to operate at higher frequencies. However, use of these higher frequencies reduced the operational effectiveness in dense foliage and urban areas.

7.2.2. Unmanned Aircraft Systems (UAS)

Small, hand-carried and/or hand-launched systems (e.g., the Raven) utilize LOS communications, while large aircraft (e.g., the Predator, Reaper, Gray Eagle, and Global

Hawk) utilize both LOS and BLOS communications, the latter generally using satellite communications.

Initial small UAS (< 20 lbs) communication systems utilize industry analog designs, but most now utilize the Army-developed digital data link (DDL) system.[21] The DDL design incorporates aspects of a software-defined radio with the ability to "field-select"[22] the frequency band in which to operate, the channel frequency within that band, the bandwidth of each channel, and the radiated power level.[23] Larger UAS operating LOS incorporate the common data link (CDL) that has been mandated for use in ISR platforms.

7.2.3. Unmanned Maritime Systems (UMS)

There are unique challenges related to UMS: dealing with the air water interface, transmission loss communicating underwater, and negotiating the dynamics of the sea surface. Intermittent communications are the norm in maritime systems and multispectral capabilities are utilized to meet communications requirements. Primary tradeoffs to be considered when evaluating a mode of communication for a USV or UUV that supports dynamic tasking, querying, and data dissemination include data rate, processing capability, range, detectability, and negotiating the maritime environment. These are of particular concern for the ISR and the ASW missions when communication is desired without exposing either the sender or receiver to possible hostile interception.[24]

7.3. Problem Statement

There are alarming red flags early in this Roadmap's time horizon regarding the amount of data that future UMS sensors will be collecting. How to best deal with that amount of data and distributing the needed information within that data to the right warfighters at the right time will be a major challenge. Left unchecked, sending all that data to local or remote sites will tax current technology and available funding (e.g., COMSAT links). The DoD needs communication technologies and tactics, techniques, and procedures that overcome these limitations and that are agile, robust, redundant, efficient, and affordable. Those needed technologies are discussed throughout Section *7.4 Way Ahead.* However, improved communication transmission technologies alone cannot achieve the necessary capacity. The DoD must pursue a fundamental shift to a future state where we pre-process the collected data, rapidly pass only critical data on to the warfighters, and then store for later retrieval the remaining data that may be needed.

In addition to achieving these technology advances [25], their application in UMS needs to meld with overall DoD wireless network communication concepts, meet national objectives for unmanned systems, and properly address regulatory policies and their limitations. All this needs to be done early in the requirements development process so those advances can be incorporated within future UMS.

In particular, tomorrow's UMS will need to utilize technical strategies which can more efficiently deal with extremely large data sets. In managing this data, better data compression, encryption and processing algorithms need to be employed in preprocessing, transmission and data fusion. These strategies also need to mandate efficient use of the spectrum, reduce frequency use overhead, allow for data security and ensure improved clarity of the available frequency spectrum. To support DoD's goals, communication systems need to support multiple frequency bands, limited bandwidth, variable modulation schemes, error correction, data encryption, and compression. All this support, of course, needs to be done so that no electromagnetic interference (EMI) [26] is caused within those systems or within other nearby spectrum-dependent systems (SDS).

There are numerous challenges to meeting this goal. First, operating a higher density of unmanned systems within relatively small areas creates increased local data rate demands. Second, size, weight, power and cooling (SWaP-C) are limiting factors on many platforms, for both onboard systems and ground/surface control systems. Third, the fidelity of the communication links must be ensured. Fourth, latency associated with digital systems must be reduced, especially for takeoff and landing of large UAS. These challenges will be exacerbated by an expected decrease in available spectrum available due to an increase in the civil[27] uses of spectrum, an objective within the Federal Communications Commission's (FCC's) National Broadband Plan[28]and directed by the White House[29]. The challenges in attaining this goal include developing, procuring, testing, and fielding communication systems that can operate with greater effectiveness, efficiency, and flexibility even in congested and adversarial environments.

Spectrum testing in unmanned systems today involves communications across a global environment with various levels of spectrum management. The communication challenges require investment in multiple technologies for leveraging communications across the radio frequencies and ultimately the optical spectrum. The impediments to unmanned systems communications are largely restricted to better use of the spectrum through investment in technologies that expand communication efficiencies. The problem today is largely a physics problem, which increases in complexity exponentially as one considers the air, ground, and maritime domain challenges. The testing of cognitive algorithms that can opportunistically leverage communications facilitating advanced mission oversight or multisystem collaboration remains in its infancy.

7.4. Way Ahead

Current DoD policies and guidance stress the need for new systems[30] to have a balance among improved interoperability; increased agility[31]; greater adaptability; improved spectral efficiency; compliance with U.S. national, host nation, and international spectrum policies[32] ; and lower production costs. The ability to update and reconfigure parts of a communication system by software changes (e.g., software-defined radios) has been available for several years. In addition, these systems should conform to a standards-based architecture (e.g., service-orientated architecture) that supports multiple networks to enable rapid and transparent configuration changes without removing the radios from operation. Such multiple-input, multiple-output (MIMO), multicarrier, and multiwaveform capabilities, along with the software control of these functions, are needed within future subsystem developments. Ultimately, it is desired that these reconfiguration changes be done "automatically" so the systems adapt dynamically (Figure 11. Autonomy Roadmap.) in response to sensed changes in the operational environment[33] (> 2020).

The need to support operations in which there are intermittent wireless propagation links has become common place. This support has resulted in increased use of advanced error control coding, MIMO configurations, various path diversity techniques, integrated networking, and data diversity — all to provide improved end-to-end quality of service. Future effectiveness of unmanned communication systems is contingent on continued advancements in antennas, transmit/receive systems, underwater communications, spectrum considerations, signal processing, network systems, and optical communications. A description of those advancements is given in the following subsections.

> ... Reinvigorate the industry's independent research and development and protect the defense industrial base.
>
> –*Under Secretary of Defense Memorandum for Acquisition Professionals, Better Buying Power, September 2010*

7.4.1. Antennas

Communication with highly mobile systems requires high-gain, rugged, and lower cost multidirectional antennas. The larger UAS systems may also use highly focused beams to achieve connectivity with more distant systems.[34] Developments in phased array antennas and "smart" antennas (to include combining signals from multiple antennas) could offer an alternative to traditional dish antennas; however, they require tradeoffs among SWaP-C. DoD and industry will need to continue developing such techniques as multifocused and super-cooled antenna systems. The multi-focused systems would permit multiple users to receive information and not rely on point-to-point systems and subsequent relaying of data via other communication systems to local users.

Future antenna systems need to be able to send and receive signals over a broad range of frequencies. Phased arrays are a viable approach. For SWaP-C and low-profile considerations, phased array antennas need to be conformal (e.g., using metamaterial) that will be molded within the vehicle surfaces. The utilization of common apertures has called for the development of new interference mitigation methodologies that minimize co-site interference effects and improve the potential for achieving simultaneous transmit and receive operations within adjacent frequency bands.

7.4.2. Transmitter/Receiver Systems

Current transmitter solid-state power amplifiers (SSPAs) are typically made with gallium arsenide (GaAs) substrate. Gallium nitride (GaN) SSPAs, currently in development, provide significant advantages over GaAs SSPAs. They offer more than double the efficiency of GaAs amplifiers; they increase the amplifier operational bandwidth; and GaN SSPAs may provide for a wider range of frequency of operation. The high transmit efficiency of GaN systems will also reduce the cooling requirements. In order to achieve some of these benefits, the amplifier designs are being enhanced with adaptive operating point control that adjusts to the instantaneous power being demanded from the amplifier. This enhancement significantly reduces the average prime power required by the amplifier by allowing it to effectively turn itself off when not in use, yet adjusting to maintain proper conditions to ensure minimal distortion at higher instantaneous powers. The GaN technologies are currently available for selected frequency bands and will soon be available for fielding (2014). The amplifiers may also utilize signal-processing-based signal predistortion techniques to compensate for the basic nonlinearity of the amplifier's transfer characteristics.

Instantaneous bandwidth performance and analog-to-digital converter sampling speeds have continued to improve year after year.[35] In addition, improvements in integrated chip fabrication methods have allowed for significant miniaturization and reductions in part counts and for various transmit/receive and antenna functions and components to be integrated on a single chip (2013). Fiber optics has been used to speed up the data and signal transfers from and to the antenna and the signal processing hardware (2012).[36] Microminiature mechanical device developments should provide smaller size, more flexibility, and greater performance in receiver designs[37] (2015). Future developments are expected to provide improvements in reliability and fabrication yields, reduced thermal characteristics, reduced integration complexity, and lower production costs.

7.4.3. Underwater Communications

Ocean dynamics challenge underwater and surface communications and are unique to UUVs and USVs. These systems gain efficiency and effectiveness with real-time, two-way communications that do not undermine mission accomplishment. The Navy's *Undersea Dominance Roadmap* (under development) will identify current and future architectures to link UUVs, distributed netted systems, and tactical platforms. Future developments described in that roadmap will leverage existing technologies and potential new capabilities that will come through the Office of Naval Research S&T research and development efforts.

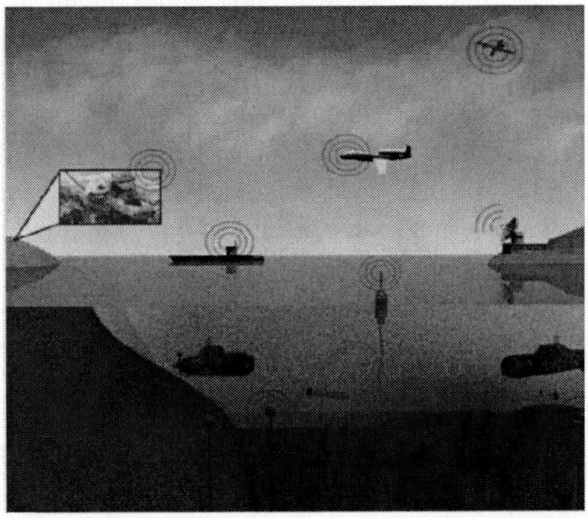

7.4.4. Spectrum Considerations

U.S. military operations are now occurring in many parts of the world where adequate spectrum is not available. There is a significant increase in the numbers of SDS being deployed by the United States, our partners, and our coalition forces to address current and expected future mission areas. In addition, these SDS collect more information, and missions often require greater bandwidths to send their information directly to warfighters. This latter consideration has been seen within OEF missions where new ISR UAS have included wide area surveillance sensors; alternative spectrum bands have been identified[38] to help address the wider bandwidths needed by those systems. Also, mission areas are becoming more spectrally "noisy" because of increasingly cluttered and hostile spectrum environments. As such, a continual demand for improved spectrum efficiency and effectiveness is being placed on all DoD SDS.[39] All unmanned systems must complete during their development process a spectrum supportability and risk assessment in accordance with DODI 4650.01 to identify and mitigate regulatory, technical, and operational spectrum supportability. Because national and international spectrum rules and policies can rapidly change,[40] developers should maintain a close liaison with appropriate DoD spectrum offices before finalizing communication system designs.

The use of LOS datalinks also supports missions where there is a denial of or impaired service to SATCOM systems. Under such conditions, the demand for LOS spectrum will be extended to support the need for improved spectrum use efficiency and effectiveness.

The Defense Advanced Research Projects Agency's (DARPA's) Next Generation (XG) project and its follow-on Wireless Network after Next (WNaN) program demonstrated the feasibility of dynamic spectrum access (DSA). DSA offers the ability to change frequency band use based on other adjacent SDS actual use and nonuse of certain bands. The Joint Tactical Radio System (JTRS) program is investigating the feasibility of integrating DSA technologies into its system. The U.S. Army is also considering having WNaN become part of an Army program of record. However, a recent USAF Scientific Advisory Board study said that DSA is far from being proven technology. Developmental challenges include susceptibility to countermeasures, costs of integrating with existing systems, developing standards (including regulatory aspects), and co-site interference (2015).

Alternative technology advances should aid in the spectral efficiency challenge to include internal and external EMI mitigation advances such as coherent signal cancellation, space-time adaptive processing, polarization diversity, and adaptive digital beam forming.

7.4.5. Communications and Signal Processing

7.4.5.1. RF Waveforms

All ISR systems are to use the CDL waveform specification[41] whenever possible.[42] A mini-CDL system[43] is finishing development to allow CDL usage within smaller platforms than were possible in the past. Candidate future CDL waveform developments include adding a "dial-arate" capability for transmission speeds (with multiple bits-per-hertz and operating within the gigabit-per-second range) and a more efficient forward error correction (FEC) coding (both 2014). Also, several UAS program offices are pursuing such performance advances as more efficient CDL waveforms, operation in additional frequency bands, expanded communications security,[44] low probability of intercept (LPI) or low probability of detection (LPD), improved anti-jam, and greater link throughput. Corresponding improvements in surface and ground stations that receive the CDL signals have also been or are being made. There have also been significant efforts to improve commonality of these systems through the UAS Task Force and its I-IPT.[45]

Future CDL improvements will include, as feasible, the incorporation of advancements being offered by DARPA, industry, and academia. Four of the most appropriate DARPA programs that are being closely followed are the DARPA Interference Multiple Access and Disruption Tolerant Networking programs.

7.4.5.2. Preprocessing

Considering the anticipated large amounts of data projected to be collected by unmanned systems in an environment of limited bandwidth capability, one challenge is determining how much of that data needs to be sent back in near real time to a ground station. There is ongoing interest in addressing how best to select portions of images and to track those portions over time and send back "just" those selected images in near real time. These selection activities are currently being developed within classified TPED programs. As these activities progress, they should be applied in preprocessing efforts onboard unmanned systems. Improved preprocessing must be accompanied by sensor and processor miniaturization to reduce SWaP-C so as to maintain the persistent nature of UMS.

FUTURE DOD R&D EFFORTS

Increased video bit depth
Quality, spatial, and temporal scalability of video
Increased number of color components
New delivery paradigms (e.g., region of interest, tiling, client/server)
Error concealment and robustness to transmission errors
Reliability of a compressed stream within noisy or high-loss communications environments
Improved compression efficiency

7.4.5.3. Compression

Compression techniques have tremendous potential to reduce bandwidth requirements, resulting in lower operating costs and increased operational flexibility. For example, FMV, synthetic aperture radar (SAR), inverted SAR (ISAR), and multispectral images can generate high bandwidth requirements (> 360 Mbit/s data rates). When compressed, the datalink bandwidth requirement could be in the range of only 1 to 30 Mbit/s.[46] Current compression techniques are described in the motion imagery systems matrix (MISM).[47] This matrix defines a recommended practice for the simple identification of broad categories of motion imagery systems. The intent of the MISM is to give user communities an easy-to-use, common shorthand reference language to describe the fundamental technical capabilities of DoD/Intelligence Community (IC)/National System for Geospatial Intelligence (NSG) motion imagery systems. The video quality needed for unmanned systems would nominally be MISM levels 4M/4H and 3M/3H. Currently the H.264 standard, which is firmly engrained in the commercial market,[48] offers twice the performance as Motion Picture Experts Group-2 standard (MPEG-2), and advanced encoding options will give even greater improvements[49] (albeit with potentially increased encoder latency).[50] The goal of the United Nations' ITU is for the H.265 standard to provide a chosen quality level at half the bit rate of H.264 (2018). For unmanned applications, future research and development should be undertaken by DoD and industry within areas depicted in the graphic on the right.

Beyond technical compression of all the collected data, there are logical advances that could reduce the amount of information that needs to be sent. This would include incorporation of logical bases for "just" replacing old information about a target's position with a more recent update, but not resending the unchanging background around the target.

In addition to performance enhancements, compression techniques have tremendous potential to reduce bandwidth requirements and thereby reduce operating costs.

7.4.5.4. Encryption

Unmanned systems incorporation of data encryption includes National Security Agency (NSA) Type 1 (for protection of classified and unclassified information) or Federal

Information Processing Standards (FIPS) Publication 140-2 certified solutions (for sensitive but unclassified information).[51] Several encryption solutions exist (e.g., Type 1 systems) for protection of unmanned systems communications (see DODI 4660). Numerous other policies and initiatives are under development within the NSA to significantly streamline the certification processes and reduce costs.[52] Future encryption solutions (2015) will inherently contain Suite B (public) encryption algorithms[53] to allow for secure classified information sharing with coalition and friendly forces. Additionally, an increasing number of encryption solutions will be based on such concepts as open standards for remote management; dynamic group keying (to support machine-to-machine information exchanges), common radio and system agnostic cryptographic interfaces (e.g., improving cryptographic component reuse and portability); software-based solutions for protection of classified data;[54] multifunctional single-chip data-in-transit and data-at-rest encryption; and single-chip all-encapsulated encryption modules (e.g., encrypt/decrypt/random key generation/key management).

7.4.5.5. Multiple Input, Multiple Output (MIMO) Systems

MIMO is a proven technology and is currently being used in commercial fourth generation (4G) wireless systems which have standards calling for a minimum of 100 Mbps for train and car speeds and 1 Gbps for stationary and walking speed. [55] MIMO combines information theory, FEC coding, signal processing, propagation theory, and consequently the mathematics behind MIMO and space-time coding is complicated. MIMO would utilize multiple paths (although not necessarily independent) with lower data rates on each path; apply space-time coding and capacity optimization to achieve a total high data rate mission; apply power saving to jammer margin; and evaluate performance in benign and stressed conditions.

With further improvements in E-discovery, interface design, and adaptive protocols, self-forming and self-healing mesh networks may enable unmanned systems to operate in multi-platform, multi-sensor type networks.

7.4.5.6. Protected Communications

In general, unmanned systems have been predominantly operated in benign environments. However, efforts are addressing improvements that are required to enable such systems to have assured and secure communications when operating in contested environments. These efforts leverage LPI, LPD, and Anti Jamming (AJ) activities that are underway in other communication systems developments. When moving UMS operations into contentious environments, a classified System Threat Assessment Report needs to be developed such that the appropriate LPI, LPD and AJ techniques are selected for incorporation into the system's design. LPD generally seeks to hide specific mission activities and involves techniques such as low power, spread spectrum, pulsed transmissions and/or directional antennas. Certain aspects of DSA could also benefit LPD. A key technique for LPI is the use of bit cover sequences within waveforms. AJ techniques include incorporating randomization at the protocol level and frequency hopping. Some aspects of DSA software implementation could offer some AJ protection.

7.4.6. Network Systems

Networking of multiple unmanned systems may be necessary to better ensure connectivity of the systems in non-LOS, urban, hostile, and/or noisy EMS environments to relay or transfer the collected information. One such concept under development is within the DARPA's LANdroids program,[56] which calls for the deployment of small, inexpensive, smart robotic radio network relay nodes that can leverage their mobility to coordinate and move autonomously. It seeks to demonstrate the capabilities of self-configuration, self-optimization, self-healing, tethering, and power management. Another concept would be the application of service-orientated architecture approaches to future network configurations.

7.4.7. Optical Communications

The application of lasers in unmanned systems communications could provide increased target detection capabilities, improved anti-jam performance, and decreased EMI within the communication subsystem. Optical communication systems are hampered by atmospheric absorption challenges, yet they offer far greater bandwidth (measured in gigabits-per-second) capabilities. LOS optical links have been successfully demonstrated at link ranges in excess of 50 km. Applications could apply to fixed locations and in air-to-air and ship-to-ship scenarios. Theoretical estimates indicate that air-to-ground links are feasible at rates up to 100 Mbit/s for link slant ranges up to 100 km, depending upon atmospheric conditions. Due to the extreme narrow beamwidth of such systems, maintaining pointing accuracy to and from a moving unmanned system will be a major challenge (> 2020).

7.5. Future Trends

Based on the force multiplier that unmanned systems have provided to our combat troops, it is expected that there will be a continued and increasing demand for supported capabilities communication systems. Those demands will include such capabilities as a single operator conducting more real-time analysis of multiple situations, while the unmanned system performs many of its assigned functions autonomously. Future communications equipment will need to be simple plug-and-play payloads that are easily, quickly, and cost-effectively modified, updated, and/or upgraded.

Future communications equipment will need to be simple plug-and-play payloads that are easily, quickly, and cost-effectively modified, updated, and/or upgraded.

7.6. Summary

There is tremendous worldwide competition for a finite amount of bandwidth. Concurrently, there is an increased demand for our unmanned systems to provide greater resolution, more persistent coverage, and continuous information flow. Technology supporting physical and software advances, and a fundamental shift in how we process and move vast quantities of data must be used to help overcome these conflicting requirements. Figure 15 provides a glimpse into the future capability and technologies we can expect throughout the course of this Roadmap.

8. TRAINING

8.1. Functional Description

Training is a critical link in delivering warfighter capability. DoD can acquire and deliver the most technologically advanced piece of machinery, but if the operators, maintainers, and support personnel are not properly trained, here is no warfighting capability. The criticality of this fact is emphasized by the requirement for acquisition program managers to "work with the training community to develop options for individual, collective, and joint training" as part of the acquisition process.[57]

Training is a learning process that involves the acquisition of knowledge, sharpening of skills, concepts and rules, or changing of attitudes and behaviors to enhance performance. Once initial training is complete, proficiency is maintained through continuation and Joint training. Unmanned systems present a unique training challenge due to the following factors:

- Availability of training areas/NAS integration.
- Frequency spectrum management.
- The rapid proliferation of numbers and types of unmanned systems in response to wartime demand.
- Differing organizational perspectives on vehicle operator qualifications, sensor operator qualifications, and support personnel requirements across the growing number of systems in all classes of UAS, UGS, and UMS.
- The reality that most USAF day-to-day continuation training is accomplished under in-theater combat conditions due to the high demand for UAS assets in real-world contingencies. This trend provides limited opportunity for the USAF to integrate its unmanned systems in pre-deployment training as the Army and USMC routinely do.
- Lack of operator interoperability and universal design standards for unmanned systems control stations.
- Lack of formalized joint tactics across the Services.

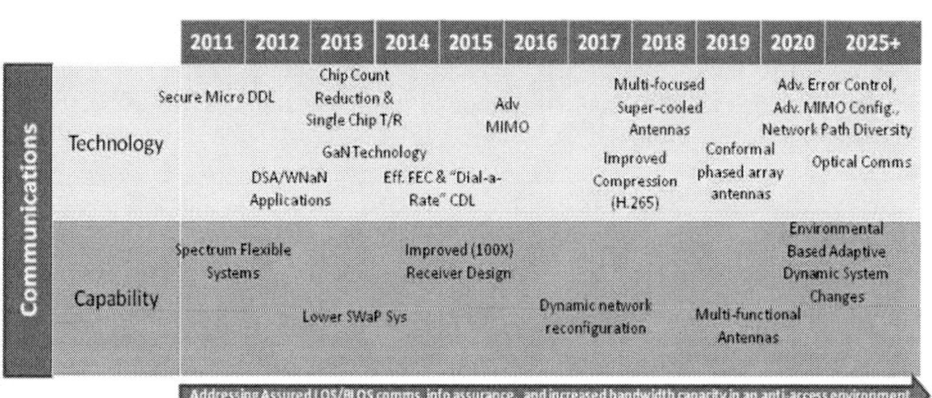

Figure 15. Communications Roadmap.

8.2. Today's State

As unmanned systems have matured and acquisition programs of record have emerged in all Services, a concerted effort has been made to ensure, wherever practical and possible, that the Services share logistics costs and burdens to include training and training systems. To date, many success stories can serve as a template for moving toward the vision of maximum joint training standard(s) for unmanned systems:

- USMC and Army personnel operate a joint Shadow UAS qualification course at the Army's training facility in Fort Huachuca, Arizona. While the Navy does not operate Shadows, Navy operators and maintainers were asked to help bridge a high-priority capabilities gap. This task was accomplished with no change in training hardware and software, simulation, or practical hands-on training. Navy personnel successfully deployed the Shadow system.
- The Chief of Staff of the Air Force and the Chief of Naval Operations signed a MoA to better utilize joint efficiencies in the Air Force Global Hawk and the Navy BAMS UAS programs. The goals of the working group are transparency between systems and a common work environment for both USAF and Navy operators.
- Army and Navy/USMC personnel share Raven B training and equipment, including maintenance requirements and GCSs.
- The JUAS COE[58] developed joint CONOPS, training qualification standards, and Tactics, Techniques, and Procedures for UAS.

Despite these success stories, given the DoD mandate to maximize training procedures and standardization for unmanned systems, the current state of unmanned systems training is still very much a work in progress.

The need for a comprehensive UAS training strategy was highlighted in UAS training workshops held in July and November 2009, hosted by the Office of the Deputy Assistant Secretary of Defense for Readiness, Directorate for Training Readiness and Strategy (ODASD(R)TR&S). The workshops were attended by all four Services, CCDRs, OSD, and Joint organizations involved with UAS issues. Additionally, a recent Government Accountability Office report recognized the lack of UAS training planning and called for the development of a DOD results-oriented strategy to resolve challenges that affect the ability to train personnel for UAS operations.[59]

> Provide soldiers and leaders the ability to excel in a challenging and increasingly complex future operating environment by developing tools and technologies that enable more efficient and effective training through live, virtual, constructive and mixed venues. Future training must enable the future force to impart more skills, faster, at lower cost and with greater retention than currently achievable. Soldiers and units must be able to be trained using non-traditional home station training techniques and technology and train prior to employment. Future training must enhance and account for individual proficiencies and learning rates (i.e. outcome

based training). Future training and leader development must be completely adaptable and scalable to cover the full spectrum of operational challenges facing the Soldier.

– Capability Gap/Deficiencies, Robotic Systems Joint Project Office Unmanned Ground Systems Roadmap, July 2009

8.3. Problem Statement

As forces drawdown in theater and redeploy, the Services will require comprehensive continuation and Joint-forces training in the peacetime environment at beddown and selected Joint-training locations. Failure to prepare for this eventuality will result in a loss of combat gained experience.

8.4. Way Ahead

The ODASD(R)TRS is leading efforts to develop a comprehensive DoD UAS training strategy. The strategy will leverage the skills and expertise of each organization and build on foundational efforts already completed or underway within the Services. The study will investigate and assess the adequacy of existing and forecast joint, Service, and CCDR UAS plans and programs that identify and describe qualification, continuation, and joint training requirements and CONOPS. The strategy will identify and describe individual, unit, and large force training requirements of all groups of UAS. The result will be a UAS Training Roadmap that guides UAS training shortfall and mitigation analyses, provides UAS training recommendations, and proposes investment considerations for the UAS community. The UAS Training Roadmap will serve as a companion piece to this Unmanned Systems Roadmap to provide a total look at efforts related to delivering UAS capabilities to the warfighter.

Intuitively, some issues that will need to be addressed in the future include:

Policy: As attention shifts more towards day-to-day continuation training and UAS are further integrated into the NAS, unforeseen disconnects in the ability to train will need to be addressed in policy.

Education: UAS need to be habitually integrated into the kill chain in training scenarios. Commanders must be educated on the use of UAS as combat resources, and learn how to train with these relatively new assets. Tactical, Operational, and Strategic level UAS and ISR doctrine should be included in appropriate professional military education courses of instruction. Issues involving operator currency, flight minimums, and continuation training requirements must be learned and opportunities to train must be must be emphasized during home station training, combined exercises, and Joint Combined Training Center rotations.

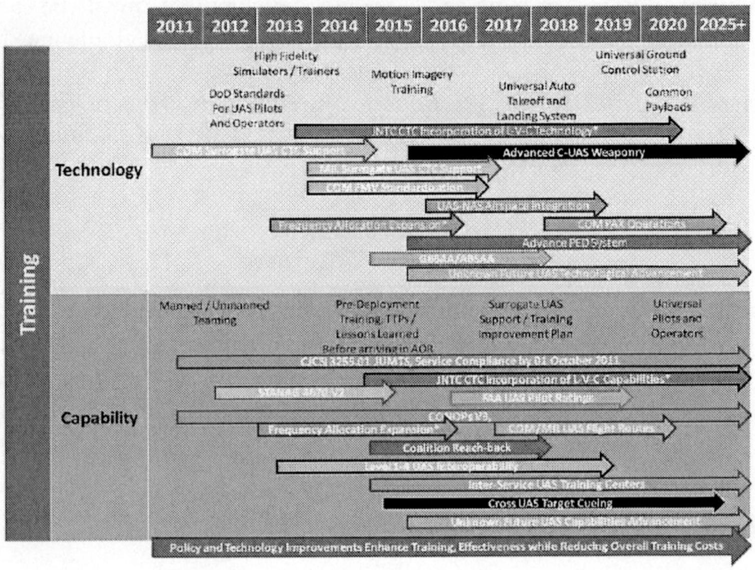

Figure 16. Training Timeline is notional. A DoD UAS Training Strategy is currently in development that will add specificity once developed.

Training Automation and Simulation: Rapidly expanding weapons systems capability requires associated expansion in training simulation. This expansion will need improved simulation fidelity and integration with live platforms for both effective/efficient use of resources. This will require improvements in training environments and classroom courseware.

Basing and Acquisition: As training requirements are defined, existing capabilities at proposed basing locations must be assessed against that which must be acquired to provide effective training.

> … UAS operators advised that the use of simulation is critical to their preparation for combat. UAS simulation is so accurate and realistic that, specifically for the Shadow UAS, it is hard to tell the difference between the simulator and actual flight.
>
> *– SFC Brian Miller, UAS Standardization NCO, Directorate of Evaluation and Standards, USAACE, Fort Rucker*

> The majority of flight training is simulation.
>
> *– SSG Brian Morton, 15W UAS Instructor/Standardization NCO, UAS Training Battalion, Fort Huachuca*

9. Propulsion and Power

9.1. Functional Description

The dramatic increase in the development and deployment of unmanned systems across the entire spectrum of air, ground, and maritime mission requirements has led to a concurrent increase in the demand for efficient, powerful, often portable, and logistically supportable solutions for unmanned system propulsion and power plant requirements.

For the purpose of this section, propulsion and power consist of the prime power to provide thrust and electrical power conversion, management, and distribution necessary for the operation of the electrically driven subsystems required to perform an unmanned vehicle's mission.

9.2. Today's State

A wide array of propulsion systems is used in unmanned systems, including combustion engines powered by heavy fuel or gasoline, jet engines, electric systems, fuel cells, solar power, and hybrid power systems. These propulsion systems can be divided into three groups according to vehicle size and mission: turbine engines, internal combustion, and electrical. The thresholds are not simple or clean cut, but are highly dependent on mission goals. Some of the parameters taken into consideration to determine the optimum propulsion system include size, weight, airflow, range, efficiency, and speed. Similarly, numerous power systems are in use, including batteries, engine-driven generators, solar power and hybrid systems. The T&E of propulsion and power is critical as we consider a world of declining energy reserves and the strategic initiatives in alternative energy being made by the DoD.

9.3. Problem Statement

Endurance is perhaps one of the most compelling aspects of unmanned systems. While power and propulsion systems are much improved over comparable manned systems, the search continues for even more efficient systems to provide greater endurance, speed and range such as the X-51A Scram Jet shown in Figure 17 preparing for first flight.

9.4. Way Ahead

9.4.1. Propulsion

> A primary long-term goal in aircraft propulsion is to reduce system specific fuel consumption by more than 30 percent over (current) gas turbine engines.... Technical challenges being pursued include efficiency, high-overall-pressure-ratio compression systems; variable-cycle engine technologies; advanced high-temperature materials and more effective turbine blade cooling; and techniques to more efficiently recuperate energy while satisfying thermal and power requirements.
>
> – *The National Plan for Aeronautics Research and Development and Related Infrastructure*

These challenges are currently being addressed for UAS applications under the highly efficient embedded turbine engine (HEETE) and efficient small-scale propulsion (ESSP) products, which are part of the Versatile Affordable Advanced Turbine Engines (VAATE) Program.

HEETE will demonstrate engine technologies that enable fuel-efficient, subsonic propulsion that supports future extreme endurance and range requirements with embedded engines incorporating complex inlets and exhausts. Covering the thrust class of 20,000 to 35,000 lbs, HEETE has two challenges: packing a high-bypass engine internally and delivering large amounts of electrical power regardless of throttle or flight condition. The HEETE design provides very small, high-powered cores to enable high bypass within the diameter constraints of an internally packaged engine. The propulsive efficiency is provided by highly efficient fans designed with the distortion tolerance needed to run behind complex inlets. The HEETE cores run at impressive pressure ratios, greater than 2.3 times the current state-of-the-art, and such ratios enable high tolerance of auxiliary power at high-altitude, long-endurance (HALE) altitudes. See Figure 18 HEETE cutaway view.

ESSP will conduct various demonstrations leading to reduced specific fuel consumption (SFC), increased power density, and a heavy fuel consumption capability. These demonstrations include a ducted fan, a nutating engine, a heavy fuel engine conversion, and a recuperator. ESSP is also designing and rig-testing high pressure ratio compressors and high temperature capable turbine concepts aimed at long-term capability.

The ducted fan is the most complex of the near-term demonstration s. The two main technologies to be demonstrated are the high-bypass geared ducted fan and the variable turbine nozzle. The test demonstrates the capability to run the high-bypass ducted fan with airflow from two different distributed core gas generators for maximum power during takeoff and maneuvering and then turning off one core gas generator, as a variable cycle feature, at cruise to cut the fuel consumption (conventional high-bypass turbofans would have to pull back the power setting to attain cruise condition, and this method would decrease engine speed, reduce the pressure ratio, and decrease component efficiencies resulting in increased SFC). The remaining core gas generator used to drive the ducted fan at cruise condition continues to operate at its design point for best cycle SFC. The variable turbine nozzle matches the airflow changes to maintain efficient turbine performance and drive the ducted fan.

Figure 17. X-51A Scram Jet.

Figure 18. Highly Efficient Embedded Turbine Engine (HEETE).

The nutating disk engine leverages small business innovation research (SBIR) contracts for both the 4-inchand 8 -inch disk engines. Both engines utilize the OSD SBIR-derived advanced microcomponents to enable engine performance potential. The major technical challenges are the development of micro-fuel injectors and radial engine seals and the understanding of the thermodynamics process. Both sizes of disk engine have undergone initial testing and show a significant increase in power density, to 1.38. The nutating disk is scalable to multiple UAV platforms by scaling the disk size.

The heavy fuel conversion engine, i.e., the Rotax used in the Predator, runs on aviation gasoline (AvGas, 100 octane). The Rotax concept demonstration is aimed at running the engine initially with lower octane fuels and ultimately with JP-8 heavy fuel. Engine testing has been completed successfully with 70 octane fuel. Although octane level is not specified for JP-8 fuel, fuel analysis to date has shown variations between a 20 to 50 octane level. Testing is on-going to demonstrate the operation of the Rotax engine on JP-8 fuel with targeted completion by the end of 2010. In parallel, there are SBIR efforts working to convert the Shadow UEL AR-741 engine to JP-8 fuel. Conversion efforts are aimed at maintaining engine performance levels while operating with JP-8 fuel.

The WTS126 turbo generator, developed by Williams International to drive the General Motors electric car, has a highly efficient recuperator, but is too heavy and large for installation into a flight vehicle. VAATE II studies indicated that a less efficient recuperator appeared to be the best balance among performance, size, and weight for a flight vehicle application. The WTS126 is an alternative heavy fuel propulsion system candidate for the Shadow. Testing and evaluation of the baseline WTS126 and the version with the less efficient flight weight recuperator are both underway.

For smaller platform applications, fuel cells offer an attractive alternative for internal combustion engines as field power generators, ground vehicle and aircraft auxiliary power units (APUs), and primary power units for small UAS. Fuel cells are devices that electrochemically combine fuel and air to produce high-quality electrical power. Because these systems do not generate power via combustion processes, they offer significantly lower SFC rates relative to advanced heavy fuel engines or diesel power generators (see Figure 20).

Solid oxide fuel cell (SOFC) systems represent a compelling power system option due to their high efficiencies, fuel flexibility, and low audible signature. Compared to other fuel cell approaches, the thermal environment and conductivity mechanism in SOFCs allow for a considerable improvement in fuel tolerance and provide a path forward for electrochemical logistic fuel operation.

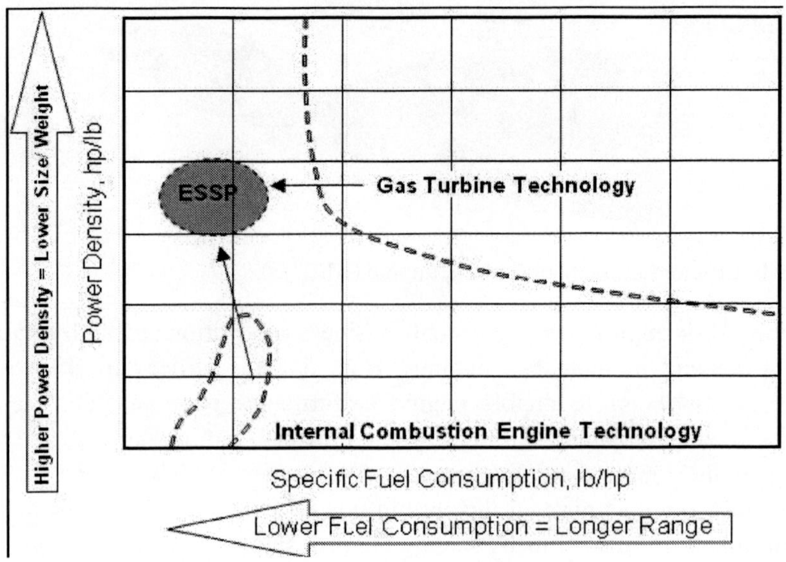

Figure 19. Efficient Small-Scale Propulsion (ESSP).

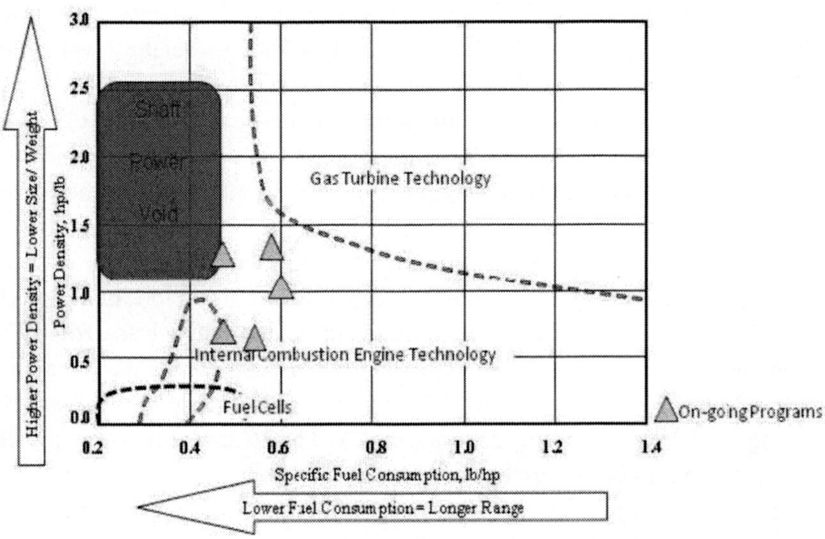

Figure 20. Fuel Cell Efficiency.

9.4.2. Power

Power sources are critical enablers for all of the desired unmanned systems capabilities. Improved power sources will have to be compact, lightweight, and reliable; provide enough power for the desired mission; and satisfy a full range of environmental and safety requirements. Design of power sources must be optimized for specific platforms and use profiles. Depending on the platform and mission requirements, applicable technologies may include energy harvesting (e.g., photovoltaic), electrical energy storage devices, fuel cells, and generators. It may be attractive to hybridize two or more of these technologies depending on the expected use profile. To implement these hybrid systems, the development of the proper control schemes must also be conducted. Recently, there has been a lot of effort invested to improve the power density of power generation systems with very good progress, but work is still needed to improve other power systems critical metrics. Some of these needed metrics and improvements are life, reliability, efficiency, optimized performance over varying engine speed, wide temperature range, production variability, control strategy, and parameters that capture the fact that unmanned subsystems typically do not have the redundancy of manned systems. Early scrutiny of the vehicle design will lead to improved power management. Form factor, materials, autonomy in sensor usage and route planning, and consideration of the undersea physical environment will minimize the energy demands and give back energy to extend the endurance or meet other mission goals.

Advances in mission equipment are providing much greater capabilities, but at a cost of greater demand for electric power, which results in greater power extraction from the engine. Power-sharing architectures allow for tailoring the source of power generation to minimize the cost in fuel burn. For example, if low-pressure (LP) power extraction is more economical than high-pressure (HP) power extraction, then the SSPCs can be turned on to power the bus that was previously powered by the HP-driven generator. Engine power extraction technologies related to power sharing between the HP spool and LP spool promise to provide significant benefit to bridging the gap between the platform power requirements and the engine power extraction limitations. Additionally, LP power extraction promises to provide improvements to SFC for overall air vehicle energy efficiency. Some of the key technologies needed to implement a power-sharing architecture are reliable power management control logics, high-power high-speed solid-state power controllers (SSPCs), a modulating generator control unit (GCU), and high-capacity electrical accumulator units (EAUs).

The HP GCU can be used to reduce the HP generator output and thus in a similar manner reduce the load on the HP spool to allow the LP generator to fulfill the power demand. The EAUs will be used to support radar peak-power demands and the power demands of short-duration, defensive-directed energy weapons.

9.4.3. Future Opportunity

Work is still needed to demonstrate the shaft power void. However, the large-engine approach of high overall pressure ratios (going to typical small-engine-corrected flow levels) is not available to small engines because of the physical size constraints of turbomachinary. Therefore, nontraditional configurations need to be emphasized to achieve the next level of capability.

Concerning battery chemistries and fuel cells, in the near term (up to 5 years), incremental power and energy performance improvements will continue to be made in the area of rechargeable lithium ion batteries. Lithium ion batteries will see broader military and commercial application, and significant cost reductions will be made as the manufacturing base matures. Near-term availability of small, JP-8 fuel-compatible engines is expected. There is midterm (5 to 15 years) potential for significant incremental performance advances through the discovery and development of alternative lithium ion chemistries. Mid-term development of fuel cells with moderate power levels (100 W class) will begin to be introduced based on low-weight hydrocarbon fuels (e.g., propane). The technical feasibility of heavy hydrocarbon-fueled (e.g., JP-8) fuel cell systems will be proven at the kilowatt class. In the long term (beyond 15 years), there is the potential for revolutionary improvements through the discovery and development of completely new battery chemistries and designs. Figure 21 charts a course for power and propulsions capabilities and technologies.

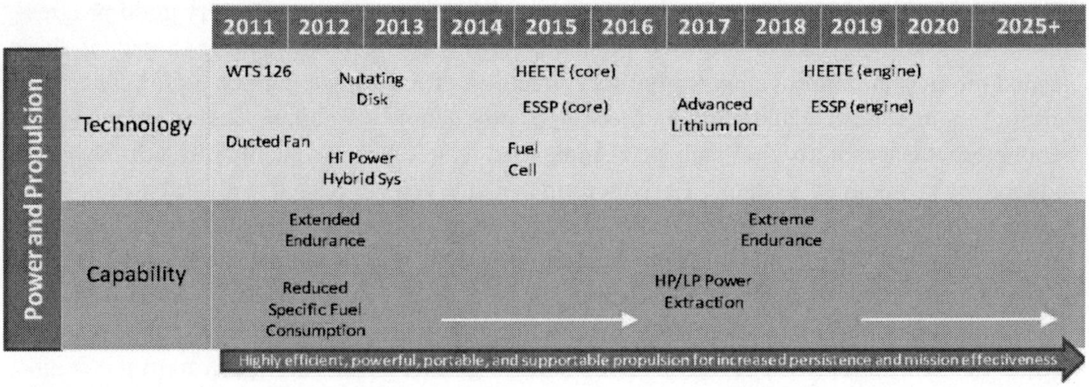

Figure 21. Propulsion and Power Roadmap.

10. Manned-Unmanned (MUM) Teaming

10.1. Functional Description

For this discussion, MUM teaming refers to the relationships established between manned and unmanned systems personnel prosecuting a common mission as an integrated team. More specifically, MUM teaming is the overarching term used to describe platform interoperability and shared asset control to achieve a common operational mission objective. This term also includes concepts of "loyal wingman" for air combat missions and segments of missions such as MUM air refueling. This capability is especially vital for missions such as target cueing and handoff between manned and unmanned systems, where the operators not only require direct voice communications between the participants, but also a high degree of geospatial fidelity to accurately depict each team member's location with regard to the object being monitored.

MUM teaming was first employed in the late 1960s when the USAF flew AQM-34 equipped with Maverick missiles from airborne C-130 aircraft. Over the intervening years, other experimental UAS were flown from manned aircraft and during the Predator ACTD from a submarine. In 2002, the USAF demonstrated the ability to fly the MQ-1 from a flying C-130 also equipped with a FMV camera to prove a rapid, small-footprint deployment capability, and the ability to cooperatively prosecute targets with onboard and off-board systems. The Army also conducted MUM demonstrations beginning with the Airborne Manned/Unmanned Systems Technology (AMUST) Demonstration in 2001 with a follow-on Hunter Standoff Killer Team (HSKT) ACTD in 2006. During that demonstration, an AH-64D executed level of interoperability (LOI) 4 control of a RQ-5B Hunter UAS during a live fire exercise where Apaches lased for their own Hellfire missiles with the Hunter payload.[60] At these demonstrations, the Army Aviation Applied Technology Directorate successfully integrated a Mobile Commander's Associate[61], including UAS control, Link 16, and other various data links, into an Army airborne C2 system. This integration enabled an airborne C2 system operator located in a UH-60 Black Hawk helicopter to control a Hunter UAS and its sensor, for the first time, as well as send and receive tactical information in flight between strike aircraft such as the FA-18, and reconnaissance aircraft such as JSTARS.[62] To date, each of the demonstrations merely changed the location of the control of the vehicle off the ground. This change was still significant because of the ability to more effectively conduct certain types of missions through collaboration of all assets.

10.2. Today's State

Practical applications of MUM teaming continue to evolve as confidence in unmanned vehicle reliability and functionality matures. Employment concepts are limited by data links, vehicle control interfaces, and level of autonomy. One recent example of practical application is when the USMC fielded a laser designation capability for Shadow as an enhancement/enabler of sensor-to-shooter operations for enemy/target of interest engagement in April 2010.

10.2.1. Unmanned Ground Vehicles (UGVs)

MUM teaming has steadily increased as technology has improved and users have found new and innovative methods to exploit this enhanced mission capability. Current missions include reconnaissance, surveillance, and target acquisition (RSTA); transport; countermining; explosive ordnance disposal; and the use of armed unmanned tactical wheeled vehicles for checkpoint security inspections. The integration of one-system remote video terminal (OSRVT) technology and distributed UGV control into ground combat vehicles is leading to the adaptation of TTPs because all parties now receive the same picture at the same time, regardless of their location.[63] With over 4,000 OSRVT or like systems fielded between the Army, USMC, and USAF to date, it is clear that MUM teaming is becoming ever more pervasive in ground operations. These developments have also been the catalyst for the creation of the common robotic controller, a joint project between the Army and USMC to develop a universal, wearable controller to operate a wide variety of unmanned systems, including UGVs, UA, and unattended ground sensors. This effort is currently aimed at smaller platforms, but could be transitioned to include limited control (i.e., payload only) for larger platforms as the technology matures.

10.2.2. Unmanned Aircraft Systems (UAS)

MUM teaming has been successfully demonstrated in combat operations to provide CCDRs with enduring surveillance of hostile activities in real/near-real time to accurately geolocate potential targets, to laser-designate targets, and to provide battle damage assessment. UAS have proven successful in performing their missions largely because they are able to remain visually and aurally undetected by hostile forces. They are providing the CCDR with critical tactical data, which are used to plan and support combat operations. When used in support of ground operations, UAS have proven invaluable in providing near-real-time intelligence to commanders engaged in combat and have directly contributed to successful mission completion. Armed UAS have the ability to engage targets directly or cooperatively with other air and ground systems. Additionally, LOI 3 (control and monitoring of the UA payload in addition to direct receipt of UA data) has been demonstrated successfully in combat operations with attack helicopter crews. The attack helicopter crew is able to see on their cockpit display the sensor outputs that give them overhead views to the target and surrounding area. This capability greatly enhances the attack helicopter crew's ability to identify, classify, and verify target locations to reduce the risk of fratricide. In September of 2010, the Army conducted an integration exercise featuring Apache helicopter pilots controlling Shadow, Hunter and Raven UAs.

The success of the exercise resulted in the inclusion of the LOI 2 and 3 UA control requirement into the AH-64, which gives the manned aircraft sensor and flight-path control and monitoring of the UA (less launch and recovery). The Apache Block III initial fielding is scheduled for 2012 and will incorporate LOI 2, 3, and 4 UA control. The AH-64 BLK III will have the capability to receive real-time UA FMV and the associated metadata (LOI 2), control the UA electro-optical/infrared (EO/IR) payload (LOI 3), and dynamically task the UA flight path (LOI 4), all from the front seat of the Apache. The initial combat operations in Afghanistan and Iraq validated the urgent need to integrate UAS capabilities with manned aircraft, specifically the attack platforms. Commanders recognized that they could dramatically reduce sensor-to-shooter times and improve situational awareness of helicopter pilots, while drastically reducing collateral damage and the potential for fratricide. They crafted an Operational Needs Statement for attack helicopter MUM teaming capability that led to a rapid prototype system for the Apache called Video from Unmanned Aircraft Systems for Interoperability Teaming – Level 2 (VUIT-2). The VUIT-2 system allows the AH-64 crew to receive video feeds from UA utilizing C-Band transmission. The Army has renamed this effort MUMT-2 and expanded it to UH-60 Black Hawk and OH-58D Kiowa Warriors.

Current MUM teaming applications are limited due to the fact the control interface currently requires a dedicated crew member to fly the UAS while another crew member flies the manned aircraft. However, some automated MUM mission segments are being developed. For example, the Navy and USAF have developed and demonstrated technology for MUM air refueling and have simulated cooperative MUM air combat missions.

10.2.3. Unmanned Maritime Systems (UMS)

MUM teaming is critical for the maritime environment. This is especially true for the undersea domain where physics prevent man from safely performing tasks to the same fidelity. There are many different aspects of MUM teaming for UMS that have been explored and implemented in various degrees: long-endurance undersea gliders send data ashore and receive human-initiated mission updates in near real-time; UUVs enable efficient port security, harbor defense, and mine clearance operations through change detection and autonomous investigation of mine-like objects; likewise, UUVs extend the footprint of manned hydrographic and bathymetric survey platforms to gather higher volumes of data while enabling people to focus on the tasks that require human oversight. Near-term enhancement, development and codification of Water Space Management/Prevention of Mutual Interference (WSM/PMI) doctrine and procedures will allow sophisticated collaboration between submarine or surface vessel operations and unmanned assets for mission accomplishment. Given the inherent challenges of the maritime environment, the future of MUM teaming will consist of multiple types of unmanned systems (UUV, USV, UAV, UGV) used collaboratively with manned platforms to collect, process, exploit, and disseminate data. An enduring and integrated net of undersea sensors partnered with USVs or UAVs for communication and controlled from a common command center will revolutionize how undersea missions are conducted by bringing transparency to an otherwise opaque battlespace. All maritime missions will benefit from reduced timelines and improved accuracy of information from which the combat commander can make engagement decisions.

10.3. Problem Statement

While strides have been made over the past decade to further enhance MUM teaming capabilities, several challenges persist that will continue to affect the amount of time it takes this technology to transition from the invention and adaptation phase to the acceptance phase.

This timing will also directly affect the development of MUM teaming TTPs, which in turn will dictate the speed of MUM teaming from CONOPS into DoD doctrine.

Some of these challenges are technical. They range from near-term issues such as the limited ability to integrate and deconflict various radio frequencies across a secure communications network, to far-term issues such as the ability of one person to control multiple UASs and UGVs simultaneously while flying his or her primary aircraft. This ability requires a high degree of hardware and software interoperability, scalable autonomy, human system interfaces (HSIs), new collaborative control algorithms, and network mission tools. The platforms must do significant levels of onboard processing to not only reduce bandwidth required, but also collaborate with other unmanned vehicles without operator input. Other technical challenges result from the need to make tradeoffs between size, weight, and power limitations on the various platforms and the desire for increased performance and capability. One of the biggest potential challenges to MUM operations stems from the Services' desire to introduce swarms (large numbers of micro-UAS operating semi-autonomously) into military operations with other manned and unmanned systems.

> "Everyone, regardless of the platform, receives the same information at the same time, leading to true interoperability; this is the Army's key goal.
>
> *-Lt. Col. Adam Hinsdale,*
> *Chief, UAS Division,*
> *Department of the Army Aviation Directorate*

Other MUM missions have different challenges including cargo, air refueling, interdiction in contested areas, electronic/network attack (EA), suppression of enemy air defenses (SEAD), and other traditional air combat missions. The ability to communicate from a highly maneuverable aircraft to a highly maneuverable future UAS will require significant advances in autonomy and HSI. This advancement can be compounded if LPI communication is needed for missions such as EA, SEAD, or control of long-dwell insect-size vehicles collecting information inside buildings.

10.4. Way Ahead (2011–2036)

Some key events will affect the future of MUM teaming over the next 25 years. As improvements in communications and sensor technologies evolve, new tactics will surely follow. For instance, it should be expected that there will be a shift away from the current reliance on video with operators incorporating other sensors (such as audio or tactile) to augment the tactical picture. Also, as commanders continue to integrate multiple manned and unmanned systems into their operations, they will soon be able to implement a "field of view" approach, similar to the "God's eye" perspective seen in many current video games. A commander will be able to view a target from multiple perspectives (i.e., UGV, UAS, or manned sensors), using multiple sensors, to obtain more robust and comprehensive situational awareness. As MUM advances, new HSI and autonomy will change the role of people in mission execution and dramatically increase their effectiveness.

The most significant advances in MUM operations will begin as Services migrate away from the current closed-loop scenario between sensor and shooter to networked systems. High endurance UAS already have mission teams geographically separated from the platform and from each other. Wide-area sensors are also changing the paradigm on STANAG 4586 LOIs and USIP development. Employing MUM segments as nodes on a larger network will change how missions are executed and will dramatically affect the combat effectiveness.

Investments in technologies such as automated air refueling, tactical data link control of maneuverable aircraft, and autonomy in the near term will enable "loyal wingman" operations. The effectiveness of air missions will not be achieved by a collection of assets, but collaboration between manned and unmanned systems within the context of a network. These nodes on the network will have scalable transparent control, not the brittle closed-loop control and inflexible autonomy algorithms used today.

The rapid growth in the OPTEMPO and demand for unmanned systems is a validation of their value to the CCDR. New concepts for the use of UAS, UGS, and UMS will result from experiences gained in combat. In the near future, it is likely that MUM teaming will be incorporated into an expanded set of operations.

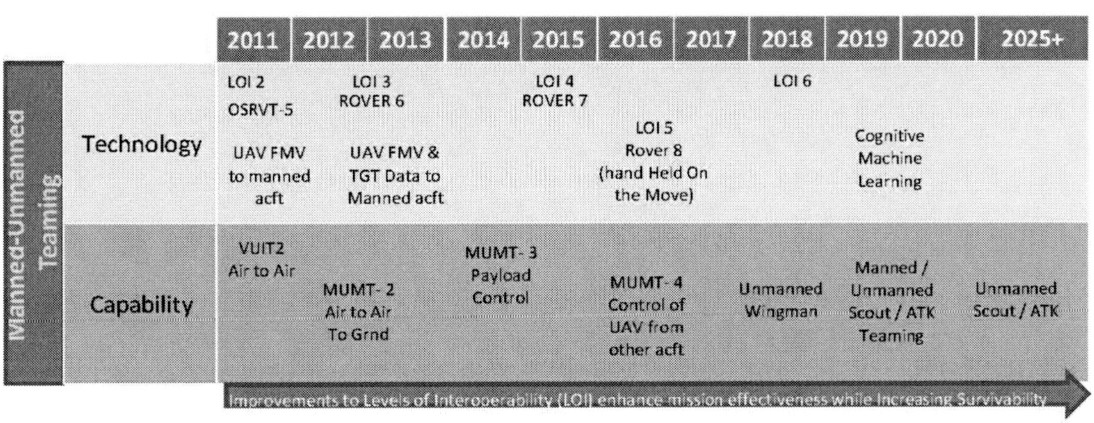

Figure 22. Manned Unmanned Teaming Roadmap.

CONCLUSION

DoD has made great strides in developing, producing, and fielding unmanned systems. These systems have been effectively integrated across air, ground, and maritime domains to support a wide range of Joint warfighting needs. The inherent advantages of unmanned systems, including persistence and reduced risk to human life, have been clearly demonstrated in combat operations in Iraq and Afghanistan.

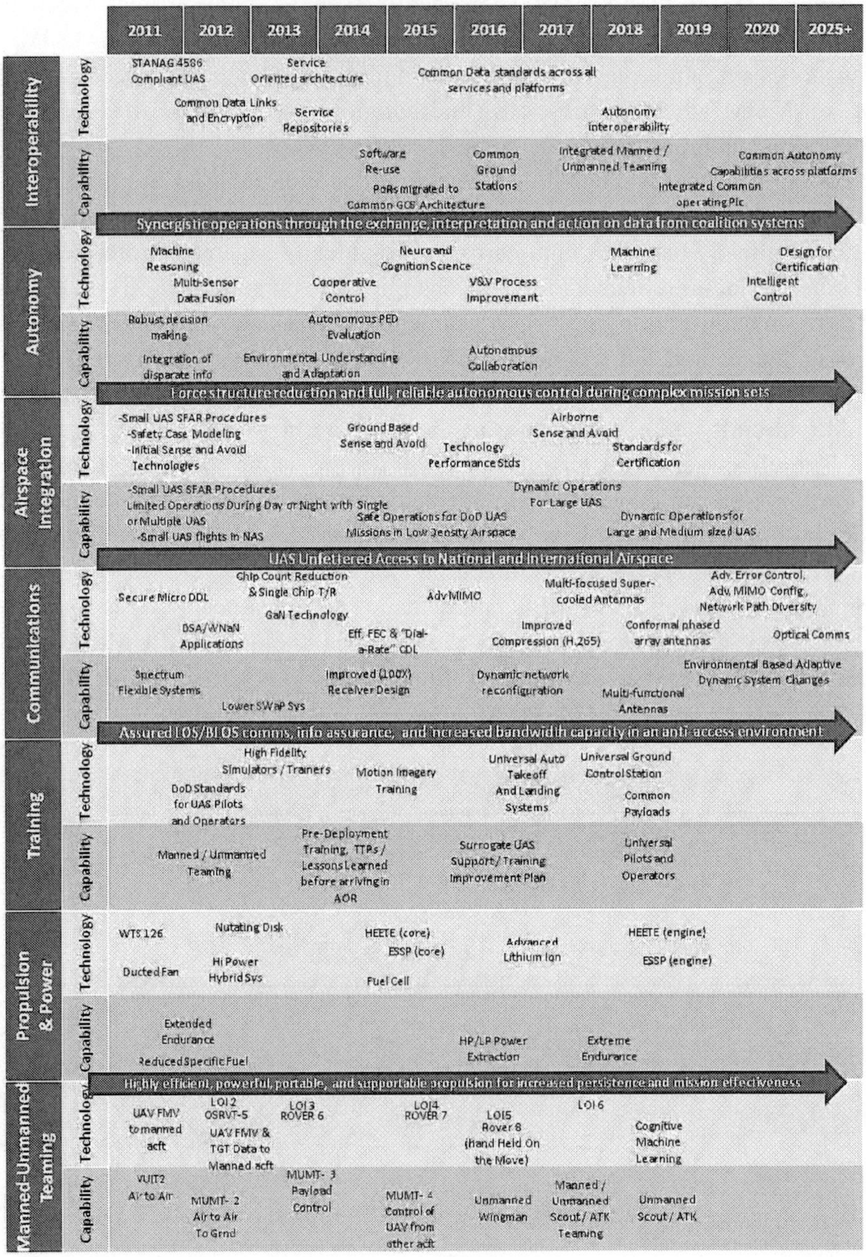

Figure 23. Summary of Technology Roadmaps.

DoD envisions the continued expansion of unmanned systems in the future force structure. This expansion will include fielding additional systems in capability areas already supported by unmanned technologies, but also expanding into new mission areas not currently covered. As DoD defines a path toward this vision, a common set of challenges is apparent that cuts across all military Services, budgets, and all three domains of air, ground, and maritime. DoD, working together with industry, academia, and other Government agencies, will continue to map an affordable path forward to address these common challenges. Success in addressing the common issues discussed in this document and following the technology roadmaps summarized in Figure 23 (see next page) is critical to achieving the full potential offered by unmanned systems technologies.

APPENDIX A. REFERENCES

[1] Unmanned Ground Systems Roadmap, Robotic Systems Joint Project Office, July 2009.
[2] Unmanned Ground Systems Roadmap, Robotic Systems Joint Project Office, Addendum, July 2010.
[3] U.S. Air Force Unmanned Aircraft Systems Flight Plan, 2009-2047, 18 May 2009.
[4] U.S. Marine Corps Concept of Operations for USMC Unmanned Aircraft Systems Family of Systems, November 2009.
[5] U.S. Navy Information Dominance Roadmap for Unmanned Systems, December 2010.
[6] Emerging Spectrum Technology Dynamic Spectrum Access Workshop Final Report, DISA, DSO, July 2009.
[7] Unmanned Aircraft System Beyond Line of Sight and Line of Sight Datalink Spectrum Technology Roadmap, UAS TF, Frequency and Bandwidth IPT, July 2010.
[8] Initial Unmanned Aircraft Systems Line of Sight and Beyond Line of Site Spectrum Findings, Interim Report, Unmanned Aircraft Systems Task Force, Frequency and Bandwidth IPT, 30 May 2008.
[9] DoD Unmanned Systems Integrated Roadmap, FY 2009-2034, April 6, 2008.
[10] U.S. Army Unmanned Aircraft Systems Roadmap, 2010-2035, 2010.
[11] Navy Unmanned Undersea Vehicle Master Plan. November 9, 2004.
[12] Joint Direct Support of Aerial Intelligence Surveillance Reconnaissance (JDSAISR) Initial Capabilities Document (ICD), 6 August 2010.
[13] U.S. Air Force Chief Scientist Report on Technology Horizons: A Vision for Air Force Science and Technology during 2010-2030, 15 May 2010.
[14] Unmanned Aircraft Systems Spectrum Report, Second Interim Report, Unmanned Aircraft Systems Task Force, Frequency and Bandwidth Integrated Product Team, September 30, 2010.

APPENDIX B. ABBREVIATIONS

4G	fourth generation
AAM	air-to-air missile

AATD	Army Aviation Applied Technology Directorate
ABSA	Aairborne sense and avoid
ACAT	Acquisition Category
ACTD	Advanced Concept Technology Demonstration
ADC	analog-to-digital converter
AECV	All Environment Capable Variant
AEODRS	Advanced Explosive Ordnance Robotic System
AI	airspace integration
AMRDEC	Aviation and Missile Research, Development and Engineering Center
AMUST	airborne manned/unmanned systems technology
APU	auxiliary power unit
ASM	air-to-surface missile
ASW	anti-submarine warfare
ATS	air traffic services
AvGas	aviation gasoline
BA	Battlespace Awareness
BAMS	broad-area maritime surveillance
C2	command and control
CA	collision avoidance
CAP	Combat Air Patrol
CBP	Customs and Border Protection
CCDR	Combatant Commander
CDL	common data link
CFR	Code of Federal Regulations
CNO	Chief of Naval Operations
COA	Certificate of Waiver or Authorization
CONEMP	concept of employment
CONOPS	concept(s) of operations
COP	common operational picture
COTS	commercial, off-the-shelf
CSS	combat services support
CV	cargo variant
DDL	digital data link
DHS	Department of Homeland Security
DIMA	DARPA Interference Multiple Access
DLI	data link interface
DoD	Department of Defense
DOTMLPF	doctrine, organization, training, materiel, leadership and education, personnel, and facilities
DOTMLPF-P	doctrine, organization, training, materiel, leadership and education, personnel, facilities, and policy
DPRK	Democratic People's Republic of Korea
DSA	dynamic spectrum access
DSPM	domain service portfolio management
EAU	electrical accumulator unit

EMI	electromagnetic interference
EMS	electromagnetic spectrum
ESSP	efficient small scale propulsion
EW	early warning
FA	Force Application
FAA	Federal Aviation Administration
FCC	Federal Communications Commission
FCS	Future combat system
FEC	forward error correction
FMV	full-motion video
GaAs	gallium arsenide
GaN	gallium nitride
GBSAA	ground-based sense and avoid
GCS	ground control station
GCU	generator control unit
GDP	gross domestic product
gMAV	Gasoline-powered Micro Air Vehicle
GPS	global positioning system
GUI	graphical user interface
HAIPE	High Assurance IP Encryption
HEETE	highly efficient turbine engine
HSKT	Hunter Standoff Killer Team
ICAO	International Civil Aviation Organization
IED	improvised explosive device
I-IPT	Interoperability Integrated Product Team
IPSEC	Internet Protocol Security
ISR	intelligence, surveillance, and reconnaissance
ITU	International Telecommunication Union
JAUS	Joint Architecture for Unmanned Systems
JCA	Joint Capability Area
JCGUAV	Joint Capability Group Unmanned Aerial Vehicle
JMOC	Joint Maritime Operations Center
JSIDL	JAUS Service Interface Definition Language
JTRS	Joint Tactical Radio System
JTS	JAUS Tool Set
LCS	littoral combat ship
LOS	line of sight
LRU	line replaceable unit
MALE	medium-altitude, long-endurance
MDF	multisensor data fusion
MIMO	multiple-input, multiple-output
MISM	motion imagery systems matrix
MISP	motion imagery standards profile
MOA	Memorandum of Agreement
MPG	multiple power pod gas generator
MUM	manned-unmanned

NAS	National Airspace System
NII	national information infrastructure
NSA	National Security Agency
NSRDEC	Natick Soldier Research, Development & Engineering Center
NUWC	Naval Undersea Warfare Center
OA	open architecture
ODUSD(R)RTPP	Office of the Deputy Under Secretary of Defense (Readiness), Readiness and Training Policy and Programs
OEF	Operation Enduring Freedom
OIF	Operation Iraqi Freedom
OPTEMPO	operational tempo
OSD	Office of the Secretary of Defense
OSRVT	one-system remote video terminal
OUSD(AT&L)	Office of the Under Secretary of Defense, Acquisition, Technology and Logistics
PEO(LMW)	Program Executive Officer of Littoral and Mine Warfare
QRF	quick reaction force
RFI	request for information
ROV	remotely operated vehicle
RS-JPO	Robotic Systems Joint Project Office
RSTA	reconnaissance, surveillance, and target acquisition
S&T	science and technology
SAA	sense and avoid
SAE	formerly known as the Society of Automotive Engineers, now known simply as SAE
SAR	synthetic aperture radar
SATCOM	satellite communications
SBIR	small business innovation research
SBU	sensitive but unclassified
SDO	standards development organization
SDS	spectrum-dependent system(s)
SF	special forces
SFC	specific fuel consumption
SIGNT	signals intelligence
SOA	service oriented architecture
SOCOM	Special Operations Command
SOFC	solid oxide fuel cell
SSPA	solid state power amplifier
SSPC	solid-state power controller
SSRA	spectrum supportability and risk assessment
STANAG	Standardization Agreement
STD-CDL	standard common data link
STUAS	small tactical unmanned aircraft system(s)
SuDDL	secure micro-digital datalink
SWaP-C	size, weight, power, and cooling
SWCC	special warfare combatant-craft crewman

TOC	total ownership costs
TPED	tasking, processing, exploitation, and distribution
TTP	tactics, techniques, and procedures
UA	unmanned aircraft UAS unmanned aircraft system(s)
UCAS	unmanned combat aircraft system
UGS	unmanned ground system(s)
UGV	unmanned ground vehicle
UMS	unmanned maritime system(s)
UMV	unmanned maritime vehicle
USC	United States Code
USIP	UAS System Interoperability Profiles
USV	unmanned surface vehicle
UUV	unmanned underwater vehicle
UW	unconventional warfare
VAATE	versatile affordable advanced turbine engine
VBSS	visit, board, search, and seizure
V&V	verification and validation
WSARA	Weapon Systems Acquisition Reform Act
WGS	Wideband Global SATCOM
WNaN	Wireless Network after Next
WRC	Worldwide Radio Communication Conference

Appendix C. Glossary

Analysis and Production – The ability to integrate, evaluate, and interpret information from available sources and develop intelligence products that enable situational awareness.

Battlespace Awareness – The ability to understand dispositions and intentions as well as the characteristics and conditions of the operational environment that bear on national and military decision-making.

Building Partnerships – The ability to set the conditions for interaction with partner, competitor or adversary leaders, military forces, or relevant populations by developing and presenting information and conducting activities to affect their perceptions, will, behavior, and capabilities.

Collection – The ability to obtain required information to satisfy intelligence needs.

Command and Control – The ability to exercise authority and direction by a properly designated commander or decision maker over assigned and attached forces and resources in the accomplishment of the mission.

Communicate – The ability to develop and present information to domestic audiences to improve understanding; and, to develop and present information to foreign audiences to affect their perceptions, will, behavior and capabilities to further U.S. national security or shared global security interests.

Communicate Intent and Guidance – The ability to promulgate a concise expression of the operational purpose, assessment of acceptable operational risk, and guidance to achieve the desired end state.

Decide – The ability to select a course of action informed and influenced by the understanding of the environment or a given situation.

Deployment and Distribution – The ability to plan, coordinate, synchronize, and execute force movement and sustainment tasks in support of military operations. Deployment and distribution includes the ability to strategically and operationally move forces and sustainment to the point of need and operate the Joint Deployment and Distribution Enterprise. (JL(D) JIC pg 5 and pages 14-21)

Direct – The ability to employ resources to achieve an objective.

Engagement – The ability to use kinetic and non-kinetic means in all environments to generate the desired lethal and/or non-lethal effects from all domains and the information environment.

Force Application – The ability to integrate the use of maneuver and engagement in all environments to create the effects necessary to achieve mission objectives.

Health Readiness – The ability to enhance DoD and our Nation's security by providing health support for the full range of military operations and sustaining the health of all those entrusted to our care.

Information Transport – The ability to transport information and services via assured end-to-end connectivity across the NC environment.

Intelligence, Surveillance and Reconnaissance – The ability to conduct activities to meet the intelligence needs of national and military decision-makers.

Intelligence, Surveillance and Reconnaissance Dissemination – The ability to present information and intelligence products that enable understanding of the operational environment to military and national decision-makers.

Intelligence, Surveillance and Reconnaissance Planning and Direction – The ability to synchronize and integrate the activities of collection, processing, exploitation, analysis and dissemination resources to meet information requirements of national and military decision-makers.

Kinetic Means – The ability to create effects that rely on explosives or physical momentum (i.e., of, relating to, or produced by motion).

Logistics – The ability to project and sustain a logistically ready joint force through the deliberate sharing of national and multi-national resources to effectively support operations, extend operational reach and provide the joint force commander the freedom of action necessary to meet mission objectives.

Maneuver – The ability to move to a position of advantage in all environments in order to generate or enable the generation of effects in all domains and the information environment.

Maneuver to Engage (MTE) – The ability to move to a position of advantage in all environments in order to employ force.

Maneuver to Influence (MTInfl) – The ability to move to a position of advantage in all environments in order to affect the behavior, capabilities, will, or perceptions of partner, competitor, or adversary leaders, military forces, and relevant populations.

Maneuver to Insert (MTI) – The ability to place forces at a position of advantage in all environments.

Maneuver to Secure (MTS) – The ability to control or deny (destroy, remove, contaminate, or block with obstacles) significant areas, with or without force, in the operational area whose possession or control provides either side an operational advantage.

Mitigate – The ability to minimize the effects and manage the consequence of attacks (and designated emergencies on personnel and physical assets.

Monitor – The ability to adequately observe and assess events/effects of a decision.

Net-Centric – The ability to provide a framework for full human and technical connectivity and interoperability that allows all DoD users and mission partners to share the information they need, when they need it, in a form they can understand and act on with confidence, and protects information from those who should not have it.

Non-Kinetic Means – The ability to create effects that do not rely on explosives or physical momentum. (e.g., directed energy, computer viruses/hacking, chemical, and biological).

Prevent – The ability to neutralize an imminent attack or defeat attacks on personnel (combatant/noncombatant) and physical assets.

Processing/Exploitation – The ability to transform collected information into forms suitable for further analysis or action.

Protection – The ability to prevent/mitigate adverse effects of attacks on personnel (combatant/noncombatant) and physical assets of the United States, allies and friends.

Shape – The ability to conduct activities to affect the perceptions, will, behavior, and capabilities of partner, competitor, or adversary leaders, military forces, and relevant populations to further U.S. national security or shared global security interests.

Supply – The ability to identify and select supply sources, schedule deliveries, receive, verify, and transfer product and authorize supplier payments. It includes the ability to see and manage inventory levels, capital assets, business rules, supplier networks and agreements (to include import requirements) as well as assessment of supplier performance.

Understand – The ability to individually and collectively comprehend the implications of the character, nature, or subtleties of information about the environment and situation to aid decision-making.

End Notes

[1] Better Buying Power, Guidance for Obtaining Greater Efficiency and Productivity in Defense Spending, OUSD(AT&L) Memo, Dr. Ashton B. Carter, 14 September 2010.

[2] http://www.dtic.mil/futurejointwarfare

[3] Updated 30 June 2011.

[4] The Combined Joint Operations from the Sea Centre of Excellence (CJOS/COE) Study (2009) for Maritime Unmanned Systems (MUS) in NATO, 8 December 2009.

[5] Consistent with Navy USV Master Plan, 2007.

[6] Ackerman, Spencer, Navy Chief Presses Nerds to Power Up Undersea Drones, Danger Room, Wired.com,8 November 2010.

[7] Definition found in Joint Publication (JP) 1-02, Department of Defense Dictionary of Military and AssociatedTerms, 12 April 2001 (as amended through 17 March 2009).

[8] DODD 5000.1, Enclosure 1, paragraph E1.10.

[9] JCOE is being disbanded June 2011 and its tasks are being transferred to the Joint Staff.

[10] Terms and Definitions, Defense Acquisition University, https://acc.dau.mil/Community Browser.aspx?id=22108.

[11] OUSD(AT&L) Acquisition Decision Memorandum, 11 February 2009.

[12] USD AT&L *Memo to Chairman, Defense Science Board, Subj Terms of Reference*, 29 March 2010.

[13] NATO Industrial Advisory Group, Study Group 75, Annex C -Autonomous Operations, 2004.

[14] Fontaine, Scott, "Schwartz outlines possible future changes," *Air Force Times*, 30 August 2010.

[15] Norman, D. A., "How might people interact with agents?" *Software Agents*, J. M. Bradshaw, Ed. Cambridge, MA: The AAAI Press/The MIT Press, 1997, pp. 49–55.

[16] Title 10 provisions relating to service authority to organize, train, and equip include 10 U.S.C. Sec. 8062 (AirForce), 10 U.S.C. Sec. 3062 (Army), 10 U.S.C. Sec. 5062 (Navy), and 10 U.S.C. Sec. 5063 (Marine Corps). Multiple service instructions address airworthiness standards, e.g., Air Force Instruction 62-601, dated 11 June2010.

[17] MIL-HDBK-516B with change 1, *Airworthiness Certification Criteria*, 29 February 2008.

[18] NATO STANAG 4671, Unmanned Aerial Vehicle Systems Air Worthiness Requirements (USAR).

[19] For the U.S. Government, see the National Telecommunications and Information Administration's *Manual of Regulations and Procedures for Federal Radio Frequency Management*. Washington, DC, January 2008 edition, September 2009 revision (incorporated by reference under 47 CFR 300.1).

[20] International Telecommunication Union (ITU), Radio Regulations, Geneva, Switzerland. 2007 Edition.

[21] Developed by the Army's Natick Soldier Research, Development & Engineering Center. An alternative DDL for small UAS is also being developed that could use a version of CDL. NATO has developed STANAG 4660, Interoperable Command and Control Data Link for Unmanned Systems (IC2DL), which is based on the DDL.

[22] This selection process is not by software but by switches. Moving to software control is being considered during a future upgrade of the DDL.

[23] The USAF Cryptographic Modernization Program Office and the Army developed a prototype encryptor for this DDL.

[24] U.S. Navy Undersea Dominance Roadmap.

[25] These advances need to include government ownership of critical data and intellectual property to ensure the best return on our research investment.

[26] The development of the resulting on-board and ground stations needs to address EMI hardening and the units tested per MILSTD-464A and MILSTD-461F.
[27] Both in CONUS and OCONUS.
[28] FCC's National Broadband Plan, Washington, DC, 2010.
[29] Presidential Memorandum "Unleashing the Wireless Broadband Revolution." June 28, 2010,
[30] This goal includes systems used in networking, communications, electronic warfare, navigation, intelligence, and sensors.
[31] This goal would include assured and secure communications.
[32] See DOD Instruction (DODI) 4650.01, Management and Use of the Electromagnetic Spectrum, Washington, DC, 9 January 2009, and DODI 4630.8, Procedures for Interoperability and Supportability of Information Technology(IT) and National Security Systems (NSS).
[33] Also see Section 5 of this Roadmap.
[34] Global positioning system (GPS) has been used to aid in this connectivity.
[35] Lundberg, Kent H., *High-Speed Analog-to-Digital Converter Survey*, MIT Press, 2002.
[36] See the DARPA Optical RF Communications Adjunct and the Office of Naval Research's Enabling Capability programs. This application is more for ground-based systems than for airborne systems. This use also significantly minimizes the signal loss and allows more advantageous placement of selected components.
[37] C. T.-C. Nguyen, "Microelectromechanical devices for wireless communications (invited)," *Proceedings*, 1998 IEEE International Micro Electro Mechanical Systems Workshop, Heidelberg, Germany, Jan. 25-29, 1998, pp. 1-7.
[38] OASD NII memo dated March 22, 2011 Subject: Department of Defense (DoD) Common Data Link (CDL) Usage in Operation Enduring Freedom (OEF) Theater
[39] All new and modified SDS programs now need to conduct a spectrum supportability and risk assessment prior to Milestone B (source: DODI 4650.01).
[40] Relatively near-term spectrum usage changes could come from the ITU and its 2011 Worldwide Radiocommunication Conference (WRC); UAS spectrum use is a conference agenda item. Changes in frequency band usage for UAS may also come from the FAA and the ICAO as part of the UAS operations in the NAS airspace and in other nation-states' airspace.
[41] These developments follow DODI 4630.09, DoD Wireless Communications Waveform Development and Management. The currently approved version is Standard CDL (STD-CDL) Rev H.
[42] The main reason for nonuse would be SWaP-C issues.
[43] This effort mostly focused on SWaP-C issues.
[44] Two cryptographic solutions (one classified and one unclassified) are used currently.
[45] See Section 4 of this Roadmap.
[46] Operational needs should determine the data rate that should be sent. Commanders in the field should be encouraged to require the lowest possible resolution and other parameters that meet their needs.
[47] See Motion Imagery Standards Profile (MISP) Recommended Practice 9720d, MISM, Standard Definition MotionImagery.
[48] It is widely used within Blu-ray and digital video disk (DVD) systems.
[49] Over the past several decades, each generation of standardized video compression has provided a halving of the required bit rate for a given quality level relative to the prior generation.
[50] The latency introduced by some compression schemes can be so great that data links using such compressions cannot be utilized during such critical times as takeoffs, landings, and weapon launches.
[51] Source: Memorandum from NII, Subject: Cryptographic Methods for Protection of Unmanned Aircraft (UAS)Wireless Communications, 6 August 2003.
[52] Management Directive 17, (U) Requirements for the Pilot Implementation to Develop Information Assurance Government Off-The-Shelf (GOTS) Secret and Below (SAB) Products and Commercial Solutions for Classified.
[53] CNSSP 15, dated March 2010.
[54] Ongoing efforts by NSA/I851.
[55] The conditions in UAS applications are much different than those for commercial cell phones.
[56] Source: http://www.darpa.mil/ipto/programs/ld/ld.asp.
[57] DoD Instruction 5000.02 Operation of the Defense Acquisition System, December 8, 2008, p.61.
[58] JCOE is being disbanded June 2011 and its tasks are being transferred to the Joint Staff and UAS Task Force.
[59] GAO-10-331, UNMANNED AIRCRAFT SYSTEMS: Comprehensive Planning and a Results-Oriented Training Strategy Are Needed to Support Growing Inventories, March 2010.
[60] "Hunter Standoff Killer Team Successfully Tests Military Interoperability," 16 September 2005, http://aero-defense.ihs.com/news/2005/navy-air-systems-link-16.htm?WBCMODE= presenta.
[61] "Mobile Commander's Associate (MCA), Lockheed Martin, USA", http://defense-update.com/products/m/mca.htm.
[62] Colucci, Frank, "MUM's The Word," *Rotor & Wing Magazine*, 1 November 2004, http://www. aviationtoday. com/rw/military/attack/1817.html.

[63] Lt. Col. Adam Hinsdale, former Chief, UAS Division, Department of the Army Aviation Directorate, was quoted in October 2007: "Everyone, regardless of the platform, receives the same information at the same time, leading to true interoperability, the Army's key goal. The OSRVT is a vital component of manned/unmanned teaming, allowing all elements, air and ground, to view the same synchronized area of interest simultaneously for coordinated engagement, with either kinetic or nonkinetic effects." *UAS Video Terminal Connects Boots On The Ground To Eyes In The Sky*, by Kim Henry, Redstone Arsenal, AL, (AFNS), 9 October 2007.

In: Technology Horizons in the U.S. Military
Editors: Ethan J. Johnson and Gabrielle L. Rodriguez
ISBN 978-1-62081-364-5

Chapter 2

TECHNOLOGY HORIZONS: A VISION FOR AIR FORCE SCIENCE AND TECHNOLOGY*

United States Air Force

EXECUTIVE SUMMARY

From its inception, the Air Force has undertaken a major effort roughly once every decade to articulate a vision for the science and technology (S&T) advancements that it should undertake to achieve over the following decade to enable the capabilities that it will need to prevail. Six such S&T visions have been developed, beginning with "Toward New Horizons" in 1945 led by Theodore von Kármán for Gen Hap Arnold, through "New World Vistas" conducted in 1995.

Since completion of the latter, 15 years have passed without an updated Air Force S&T vision. "Technology Horizons" represents the next in this succession of major vision efforts conducted at the Headquarters Air Force level. In view of the far-reaching strategic changes, rapid global technological advances, and growing resource constraints over the next decade, this is an overdue effort that can help guide S&T investments to maximize their impact for maintaining Air Force technological superiority over potential adversaries.

What Is "Technology Horizons"?

"Technology Horizons" is neither a prediction of the future nor a forecast of a set of likely future scenarios. It is a rational assessment of what is credibly achievable from a technical perspective to give the Air Force capabilities that are suited for the strategic, technology, and budget environments of 2010-2030. It is visionary, but its view is informed by the strategic context in which these technology-derived capabilities will be used. It is an articulation of the "art of the possible", but is grounded in the knowledge that merely being

* This is an edited, reformatted and augmented version of a U.S. Air Force "Technology Horizons" 2010–2030, AF/ST-TR-10-01-PR, dated 15 May 2010.

possible is only a prerequisite to being practically useful. It considers the spectrum of technical possibilities, but acknowledges that budget constraints will limit the set of these that can be pursued.

It recognizes that increasingly more of the science and technology that provides the basis for future Air Force capabilities is available worldwide to be translated into potential adversary capabilities. It thus has sought to envision not only U.S. Joint and allied opportunities for using technologies, but also ways that adversary capabilities could be derived from them using entirely different concepts of operations or on the basis of entirely different warfighting constructs. It acknowledges that capabilities enabled by new technologies and associated operating concepts often introduce new vulnerabilities not envisioned in the original capability. It thus has also considered potential vulnerabilities and cross-domain interdependences that may be created by second- and third-order effects of these technology-derived capabilities.

Sources of Inputs

"Technology Horizons" received inputs from a wide range of organizations and sources. These included four working groups, one each in the air, space, and cyber domains and another that focused on cross-domain insights, that gave a broad range of subject matter expertise to this effort. Working group members were drawn from the Air Force S&T community, intelligence community, MAJCOMs, Product Centers, FFRDCs, defense industry, and academia. Further inputs were obtained from site visits, briefings, and discussions with organizations across the Air Force, the Department of Defense, federal agencies, FFRDCs, national laboratories and industry, including Air Staff and Air Force Secretariat offices. Additional use was made of perspectives in several hundred papers, reports, briefings and other sources.

Elements of the Resulting S&T Vision

The vision from "Technology Horizons" to help guide Air Force S&T over the next decade and beyond consists of the following elements:

1. Strategic Context
2. Enduring Realities
3. Overarching Themes
4. Potential Capability Areas
5. Key Technology Areas
6. Grand Challenges
7. Summary of S&T Vision
8. Implementation Plan & Recommendations

Overarching Themes for Air Force S&T

The strategic context and enduring realities identified in "Technology Horizons" lead to a set of 12 "Overarching Themes" to vector S&T in directions that can maximize capability superiority. These shifts in research emphases should be applied judiciously to guide each research area.

1. From ... Platforms	To ... Capabilities
2. From ... Manned	To ... Remote-piloted
3. From ... Fixed	To ... Agile
4. From ... Control	To ... Autonomy
5. From ... Integrated	To ... Fractionated
6. From ... Preplanned	To ... Composable
7. From ... Single-domain	To ... Cross-domain
8. From ... Permissive	To ... Contested
9. From ... Sensor	To ... Information
10. From ... Operations	To ... Dissuasion/Deterrence
11. From ... Cyber defense	To ... Cyber resilience
12. From ... Long system life	To ... Faster refresh

Potential Capability Areas, Key Technology Areas, and Grand Challenges for Air Force S&T

Based on the strategic environment, enduring realities, and overarching themes, the remaining elements of the vision from "Technology Horizons" are presented as follows to help guide the Air Force in making choices on the most essential S&T efforts that must be pursued to prepare for the environment of 2010-2030. Specifically:

- It identifies "Potential Capability Areas" (PCAs) and maps these PCAs across the Air Force service core functions to assess the range of impact they can have.
- It uses this set of PCAs to identify "Key Technology Areas" that are most essential for the Air Force to invest in over the next decade to obtain capabilities aligned with the strategic, technology and budget environments.
- It additionally defines four "Grand Challenge" problems to advance key technology areas and integrate them in system-level demonstrations of significant new capabilities.
- It presents an implementation plan that enables the elements of this vision to be put into practice for guiding Air Force S&T efforts to maximize resulting capabilities in 2030.

Major Findings from "Technology Horizons"

The single greatest theme to emerge from "Technology Horizons" is the need, opportunity, and potential to dramatically advance technologies that can allow the Air Force to gain the capability increases, manpower efficiencies, and cost reductions available through far greater use of autonomous systems in essentially all aspects of Air Force operations. Increased use of autonomy -- not only in the number of systems and processes to which autonomous control and reasoning can be applied but especially in the degree of autonomy that is reflected in these --can provide the Air Force with potentially enormous increases in its capabilities, and if implemented correctly can do so in ways that enable manpower efficiencies and cost reductions.

Achieving these gains will depend on development of entirely new methods for enabling "trust in autonomy" through verification and validation (V&V) of the near-infinite state systems that result from high levels of adaptibility and autonomy. In effect, the number of possible input states that such systems can be presented with is so large that not only is it impossible to test all of them directly, it is not even possible to test more than an insignificantly small fraction of them. Development of such systems is thus inherently unverifiable by today's methods, and as a result their operation in all but comparatively trivial applications is uncertifiable.

It is possible to develop systems having high levels of autonomy, but it is the lack of suitable V&V methods that prevents all but relatively low levels of autonomy from being certified for use. Potential adversaries, however, may be willing to field systems with far higher levels of autonomy without any need for certifiable V&V, and could gain significant capability advantages over the Air Force by doing so. Countering this asymmetric advantage will require as-yet undeveloped methods for achieving certifiably reliable V&V. The Air Force, as one the greatest potential beneficiaries of more highly adaptive and autonomous systems, must be a leader in the development of the underlying science and technology principles for V&V.

A second key finding to emerge from "Technology Horizons" is that natural human capacities are becoming increasingly mismatched to the enormous data volumes, processing capabilities, and decision speeds that technologies either offer or demand. Although humans today remain more capable than machines for many tasks, by 2030 machine capabilities will have increased to the point that humans will have become the weakest component in a wide array of systems and processes. Humans and machines will need to become far more closely coupled, through improved human-machine interfaces and by direct augmentation of human performance.

Focused research efforts over the next decade will permit significant practical instantiations of augmented human performance. These may come from increased use of autonomous systems as noted above, from improved man-machine interfaces to couple humans more closely and more intuitively with automated systems, or from direct augmentation of humans themselves. The latter includes drugs or implants to improve memory, alertness, cognition, or visual/aural acuity, as well as screening of individuals for speciality codes based on brainwave patterns or genetic correlators, or even genetic modification itself. While some such methods may appear inherently distasteful, potential adversaries may be entirely willing to make use of them.

Developing ways of using science and technology to augment human performance will become increasingly essential for gaining the benefits that many technologies can bring. Significant advances and early implementations are possible over the next decade. Such augmentation is a further means for increasing human efficiencies, allowing reduced manpower needs for the same capabilities or increased capabilities with given manpower.

A further key theme is the need to focus a greater fraction of S&T investments on research to support increased freedom of operations in contested or denied environments. Three main research areas are of particular importance: *(i)* cyber resilience, *(ii)* precision navigation and timing in GPS-denied environments, and *(iii)* electromagnetic spectrum warfare. Additionally, the study identifies further key priority areas where S&T investment will be needed over the next decade to enable essential capabilities, including processing-enabled inteligent sensors, directed energy for tactical strike/defense, persistent space situational awareness, rapidly composable small satellites, and next-generation high-efficiency gas turbine engines.

Organization of Results from "Technology Horizons"

Results are presented in three volumes. Volume 1 provides the main findings and the elements of the S&T vision for the Air Force during 2010-2030. Volume 2 provides an overview of current S&T in AFRL. Volume 3 will provide certain classified and proprietary supporting information.

Recommendations

"Technology Horizons" makes five major Recommendations for guiding Air Force S&T efforts to meet the strategic, technology, and budget challenges over the next decade and beyond:

Recommendation #1: Communicate Results from "Technology Horizons"

- Communicate "Technology Horizons" rationale, objectives, process, and key elements via briefings offered to all HAF offices, MAJCOMs, Product Centers, and AFRL.
- Build broad awareness, understanding, and support for the Air Force S&T vision from "Technology Horizons".
- Disseminate Volumes 1 and 2 of "Technology Horizons" across all relevant organizations beyond those that provided inputs to this effort.

Recommendation #2: Assess Alignment of S&T Portfolio with "Technology Horizons"

- Assess alignment of AFRL's current S&T portfolio with the broad research directions and technology focus areas outlined in "Technology Horizons".
- Identify target fraction of the total Air Force S&T portfolio to be aligned with the research directions and technology focus areas identified in "Technology Horizons".

Recommendation #3: Adjust S&T Portfolio Balance as Needed

- Identify research efforts in current S&T portfolio that must be redirected or realigned with the research directions and technology focus areas in "Technology Horizons".
- Determine which of these efforts should be realigned, redirected, or terminated to accommodate new research efforts that achieve the needed direction and emphasis.
- Define new research efforts that will be started to allow broad research directions and technology areas identified in "Technology Horizons" to be effectively achieved.
- Implement changes in the AFRL S&T portfolio to initiate new research efforts identified above, and to realign, redirect, and terminate existing efforts identified above.

Recommendation #4: Initiate Focused Research on "Grand Challenge" Problems

- Evaluate, define, and focus a set of "grand challenge" problems of sufficient scale and scope to drive major technology development efforts in key areas identified here.
- Structure each "grand challenge" to drive advances in research directions identified in "Technology Horizons" as being essential for meeting Air Force needs in 2030.
- Define specific demonstration goals for each challenge that require sets of individual technology areas to be integrated and demonstrated at the whole-system level.
- Initiate sustained research efforts in AFRL S&T portfolio as necessary to achieve each of the "grand challenge" demonstration goals.

Recommendation #5: Improve Aspects of Air Force S&T Management Process

- Obtain HAF-level endorsement of an AFRL planning construct for S&T to provide the stability needed for effective mid- and long-range development of technologies.
- Define and implement a formal process for obtaining high-level inputs from MAJCOMs and Product Centers in periodic adjustments within the AFRL S&T planning construct
- Develop and implement an informal process to obtain more frequent inputs from MAJCOMs and Product Centers for lower-levels of the AFRL S&T planning construct.

SECTION 1. OVERVIEW OF "TECHNOLOGY HORIZONS"

"Technology Horizons" was conducted over a nine-month period to provide a vision for science and technology in the Air Force over the next decade and beyond. This Section summarizes the goals, organization, and execution of this effort, and describes how the results are presented.

The U.S. Air Force today finds itself at an undeniably pivotal time in its history. It is without question the most effective and powerful air force in the world, and the only air force that can truly project global power. That position of strength was attained by organizing, training, and equipping a professional workforce for the entire range of functions essential to Joint combat and combat support, and for non-combat missions that the Service is increasingly called on to perform. Yet perhaps the single most important factor in achieving this position has been the unmatched technological advantage that the Air Force has attained over any of its competitors.

Today however, the confluence of strategic changes, worldwide technological advancements, and looming resource constraints cause some to wonder how the Air Force will maintain this position of superiority. As the spectrum of conflict has grown, so too have the demands for a wider range of capabilities across all facets of conflict. Unprecedented worldwide diffusion of technologies is giving competitors access to capabilities that may have the potential to offset important parts of Air Force technological advantages. At the same time, indications are that defense budgets in the coming decade may make it increasingly difficult to achieve the pace of science and technology investment that has provided the foundation for Air Force superiority.

1. A Vision for Air Force Science and Technology during 2010-2030

It is against this background that the Secretary of the Air Force and the Air Force Chief of Staff called for the Chief Scientist of the Air Force to conduct the "Technology Horizons" study. The tasking and Terms of Reference for the study are at the front of this report and in Appendix A. It examined technologies across the air, space, and cyber domains to develop a forward-looking assessment, on a 20-year horizon, of possible offensive and defensive capabilities and counter-capabilities of the Air Force and its potential future adversaries that could substantially alter future warfighting environments and affect future U.S. Joint capability dominance. In doing so, "Technology Horizons" provides a vision for where Air Force science and technology should be focused over the next decade to maximize the technology superiority of the U.S. Air Force.

1.1. A Rational Technical Assessment, Not a Forecast or Prediction

"Technology Horizons" is neither a prediction of the future nor a forecast of a set of likely future scenarios. Such efforts play a useful role, but forecasts or predictions tend in most cases to overestimate the pace of progress, even when technology overall is advancing at an undeniably rapid pace. Equally important, efforts to forecast or predict the future invariably miss one or more trivial events that can have a determinative effect on the future that actually occurs. Chaos theory indicates that the particular future that will occur beyond a short time horizon is the result of a confluence of many events, the product of which can be

inherently unpredictable with any useful certainty and can be influenced to leading order by seemingly incidental events.

> *"Our Air Force is at another inflection point in its history, where changes in the strategic environment, new technologies, and changes in resources combine to reshape our capabilities and to set us in new directions."*
>
> The Honorable Michael B. Donley

Secretary of the Air Force

Neither is "Technology Horizons" a fantasy from technologists describing a future in which technology will achieve seemingly boundless wonders. The inaccuracy of such descriptions has over time served more to reveal how difficult it in fact is to translate technological possibilities into practically achievable systems while satisfying numerous operational constraints that may have nothing to do with the technologies themselves.

Rather, "Technology Horizons" is a rational assessment of what is credibly achievable from a technical perspective to give the Air Force a set of capabilities that is suited for the strategic environment of 2010-2030. It identifies the key technology areas where Air Force S&T investment is most critical for providing the Service with the ability to adapt and prevail across the spectrum of conflict in the air, space, and cyber domains. It further addresses emerging cross-domain areas where entirely new threats and opportunities exist.

It is visionary, but its view is informed by the strategic context in which these technology-derived capabilities will be used. It is an articulation of the "art of the possible", but is grounded in the scientific and engineering knowledge that merely being possible is only a prerequisite to being practically useful. It considers the spectrum of technical possibilities, but acknowledges that budget constraints will limit the set of these that can be pursued.

1.2. Science & Technology for Materiel and Non-Materiel Solutions

Air Force capabilities and costs are closely interdependent. The costs of normal operations consume resources that otherwise might be available to develop greater capabilities. Manpower costs, for example, are consuming an increasingly larger fraction of the overall Air Force budget. Technology solutions that can reduce manpower requirements or increase manpower efficiencies via non-materiel means can thus have as much impact on future capabilities as can direct materiel solutions. The S&T vision articulated in "Technology Horizons" thus seeks to identify "disproportionately valuable" materiel and non-materiel solutions over the 2010-2030 time frame.

> "*While we remain resolute about the issues that remain, we can, and we must, raise our sights to focus on the longer-term vision – an Airman's vision of constant innovation in the control and exploitation of air, space, and cyberspace.*"
>
> Gen Norton B. Schwartz
> U.S. Air Force Chief of Staff

"Technology Horizons" looks at capabilities on a 20-year time horizon, but recognizes that it can take a decade to translate technologies from maturity into fielded systems. It therefore is based on an approach that considers technology developments over the next decade that can provide the basis for technology-enabled capabilities over the following decade.

This effort also recognizes that increasingly more of the science and technology that provides the basis for future Air Force capabilities is available worldwide to be translated into potential adversary capabilities. It therefore has sought to envision not only U.S. Joint and allied opportunities for using these technologies, but also ways that adversary capabilities and "technology surprise" could be derived from them using entirely different concepts of operations or on the basis of entirely different warfighting constructs.

"Technology Horizons" further recognizes that capabilities enabled by new technologies and associated operating concepts often introduce new vulnerabilities not envisioned in the original capability. It thus has sought to consider potential vulnerabilities and cross-domain interdependences that may conceivably be created by second- and third-order effects of these technology-derived capabilities.

1.3. Can we even Afford to do any of these Things?

With factors such as manpower costs, sustainment costs, and energy costs consuming increasingly more of the Air Force budget, and the outlook for defense budgets suggesting little or no real growth over the next decade, some may rightly ask whether a study such as this is even useful. If the Air Force does not have the resources to put behind the S&T focus areas emerging from this effort, then what sense does it make to identify these and develop an Air Force S&T vision around them?

At a minimum, such a view would overlook the fact that technologies themselves can help offset many of the costs noted above. For instance:

- New technologies for increased cyber resilience of Air Force networks and systems can potentially free up manpower otherwise consumed by additional cyber specialists and their training that would be needed to defend these systems.
- Technologies can provide increased trust in autonomy to enable reduced manpower requirements via flexibly autonomous systems, or equivalently autonomous systems can enable greater warfighter capabilities for the same manpower requirements.
- Technologies can enable fuel cost savings by increases in turbine engine efficiency, by advances in lightweight materials and multifunctional structures, by advanced aerodynamic concepts and technologies, and by adaptive control technologies.
- Technologies could potentially reduce manpower needs through augmentation of human performance, via implants that improve memory, alertness, cognition, and visual/aural acuity, brainwave-coupled human-machine pairings, or even screening of individuals for specialty codes based on brainwave patterns or genetic correlators.

Such examples show how an Air Force S&T vision that properly accounts for the role that budget constraints will play in the strategic environment can enable materiel and non-materiel solutions to reduce Air Force costs while expanding capabilities.

It should be further apparent that in a budget-constrained environment it is all the more essential to have a clear vision for where S&T investments should be focused. It is precisely when resources are too constrained to allow as broad a set of S&T investments as might be desirable that the Air Force needs a vision for which technology areas are most essential for it to invest in. While broader investment will be needed to provide an appropriately hedged strategy, a substantial fraction of total S&T efforts should be aligned with a vision that identifies technology areas for enabling among the greatest Air Force capabilities.

> *"Technology Horizons" provides a vision for where a substantial fraction of Air Force S&T investment should be focused to provide the greatest possible capabilities in the strategic environment, budget environment, and technology environment during 2010-2030.*

2. Lessons Learned from Prior Air Force S&T Visions

Throughout its history, beginning shortly before it became a separate Service, the Air Force has conducted a major study roughly once every ten years to develop a vision for the role that science and technology would seek to fill over the following decade. As shown by the timeline in Figure 1, six such studies have previously been conducted, beginning with "Toward New Horizons" in 1945 led by Theodore von Kármán for General Hap Arnold, and continuing through "New World Vistas" conducted in 1995 by the Scientific Advisory Board for Secretary of the Air Force Dr. Shiela Widnall and Air Force Chief of Staff Gen Ronald Fogleman.

Since completion of the latter, over 15 years have passed without a major study to develop an updated S&T vision for the Air Force. "Technology Horizons" is the next in this succession of major S&T visions conducted at the Headquarters Air Force (HAF) level. Particularly in view of the fundamental strategic changes, rapid global technological advancements, and forseeable resource constraints that define the environment over the next decade, "Technology Horizons" is an overdue effort to help guide Air Force S&T investments to include areas that will have disproportionately valuable impacts on reducing costs and maintaining capability advantage.

"Technology Horizons" seeks to help guide the science and technology efforts of the Air Force over the next decade to maximize their impact in the substantially different environment that the Air Force is now facing.

2.1. Previous Air Force Science & Technology Visions

Figure 1 shows the six previous Air Force S&T vision studies that have been conducted at the HAF level. The duration, size, composition, organization, output, and impact of these have varied widely. The following observations can be made:

- These studies have been conducted over durations ranging from nine months to, in one case, nearly two years.
- The numbers of participants involved has ranged from as few as 25 to nearly 500.
- Participation has ranged from academic-only studies to military-only studies, and some that included other government and industry participants.
- Panels have in some cases been organized around technical disciplines, and in others around capability topics, as well as other arrangements.
- Reports have ranged from concise summaries to more than a dozen volumes and total lengths in excess of 1300 pages.
- Impacts on technical directions, organization, and other aspects of Air Force S&T have ranged widely, in part depending on how responsive the studies were to the central issues of their time.

Detailed descriptions of the first five of these prior Air Force S&T vision studies are given in "Harnessing the Genie: Science and Technology Forecasting for the Air Force 1944-1986."

2.2. Key Factors that have Produced Successful S&T Visions

As the above summary suggests, and as detailed in the reference noted, there is no correlation of the impact of these studies with either the number of participants or the length of the reports they have produced. What correlations there are have principally to do with the types of participants and the way that working panels have been organized.

Studies without representation in substantial numbers from the research community have generally not been successful, though sufficient operational inputs are essential to ground the discussions. Panels organized along relatively narrow system themes appear, as might be expected, to restrict the range of ideas for discussion, and essentially preclude cross-domain insights created by interdependences among systems. Studies with panels organized around broader themes have generally proven to be more successful.

> The principal factor that has determined the impact of prior S&T visions has been the extent to which Air Force senior leadership has embraced them as a means for helping to vector the direction of future Air Force research.

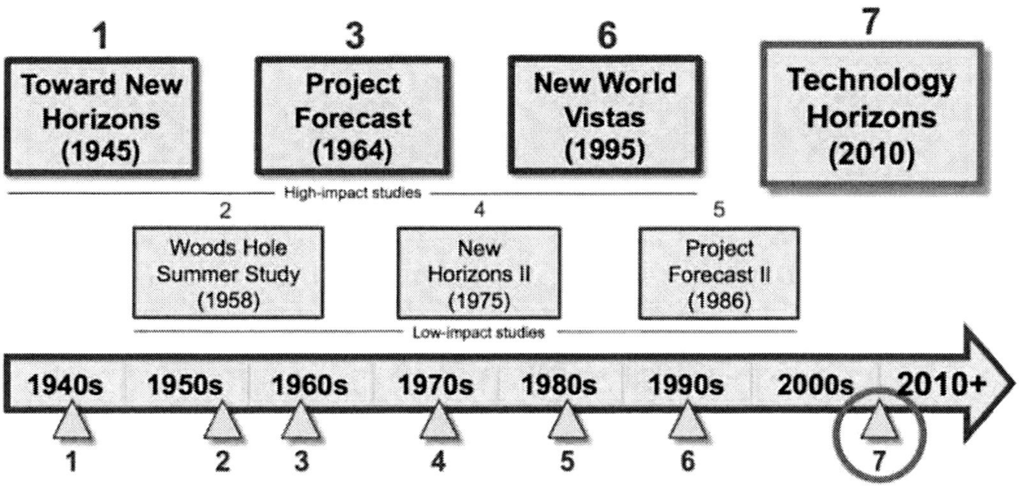

Figure 1. Summary of HAF-level studies to produce Air Force S&T visions. "Technology Horizons" is an overdue effort to produce an S&T vision appropriate for the strategic environment during 2010-2030.

3. Organization and Conduct of "Technology Horizons"

"Technology Horizons" formally began in June 2009 and was completed in February 2010. Organization of the study and the working groups preceded the formal start of the effort.

3.1. Strategic Perspective as the Foundation for an Effective S&T Vision

The Air Force S&T vision from "Technology Horizons" is predicated in part on the principle that S&T can only be usefully guided in a study like this by placing it in context with the strategic environment in which the Air Force will be operating during the 2010-2030 time frame it addresses. The technology areas that are recommended for Air Force S&T to put special emphasis on are those that meet key needs dictated by this strategic environment. "Technology Horizons" thus is not simply an opportunity-driven technology vision, but matches the demand-side pull of the Air Force during 2010-2030 with the opportunity-side push that realistically achievable technologies can enable.

3.2. Technology Horizons "10+10 Technology-to-Capability" Process

To match technology opportunities with the needs of this strategic environment, the study used a "10+10 Technology-to-Capability" process shown in Figure 2.

This recognizes that even if it were possible to project the state of *technologies* 20 years into the future to the 2030 horizon date of the study, this effort is instead designed to help the Air Force envision what technology-enabled *capabilities* it could credibly have to meet its key needs in 2030. Thus the study must determine the state to which the necessary supporting technologies can be brought in time to enable those capabilities by 2030.

Owing to the 10 years or so that it typically takes to transition TRL 6-level technologies into fielded capabilities, this means that the underlying date to which technologies must be projected is 2020. That in turn allows a determination of the technology investments that need to be underway today, in order to enable these Air Force capabilities in 2030.

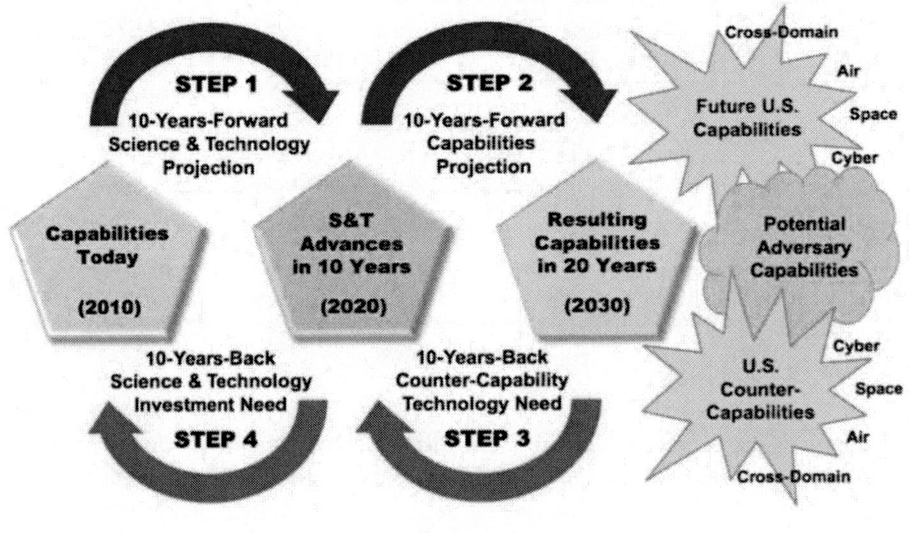

Figure 2. Schematic of the *10+10 Technology-to-Capability* process used in "Technology Horizons".

The required 10-year projection from today's state of technology can be credibly made with acceptable uncertainties, even under the rapid pace at which technologies are advancing. By contrast, the uncertainties if a 20-year projection of technologies were needed from today's state would be far greater than twice as large. Thus the 2030 time horizon of "Technology Horizons" requires only a credible projection from 2010 to 2020 of the technologies needed to enable a set of capabilities that could be fielded in 2030 to meet key needs of the Air Force in the strategic environment.

> Technology Horizons will help the Air Force obtain the greatest effectiveness in a budget-constrained environment; if we invest in the right technology areas we can have unbeatable capabilities.

A further foundation of this study is that potential adversaries will have access to much of the same science and technology as the Air Force does over this period, and thus can develop red force capabilities from these that -- driven by entirely different concepts of operations -- may be entirely different from blue force capabilities developed from them. It thus becomes essential to envision not only credible U.S. technology-derived capabilities in 2030, as shown on the right in Figure 2, but also potential adversary capabilities along with appropriate U.S. counter-capabilities, as also shown on the right in the figure.

Developing the technologies needed to enable these counter-capabilities then involves another pair of 10-year steps, in this case shown in red in Figure 2. The first of these steps begins with the needed counter-capabilities in 2030 and projects back to the required state of the underlying enabling technologies in 2020. The second of these then does a further 10-year projection back to 2010 to determine what S&T investments would need to be underway today to thereby enable these counter-capabilities in 2030.

The result is the four-step “10+10 Technology-to-Capability” process in Figure 2:

- Step 1: Beginning with the state of technologies today, a 10-year forward projection is made of the state to which these technologies, or ones derived from them, can be brought in 2020.
- Step 2: From the state to which technologies can be brought in 2020, a further 10year forward assessment is made of the U.S. capabilities that could be enabled by them to meet key needs of the strategic environment in 2030.
- Step 3: Potential adversary capabilities and needed U.S. counter-capabilities are envisioned in 2030, and a 10-year backward projection is made of the state to which technologies needed to enable these counter-capabilities must be brought by 2020.
- Step 4: From the required state of technologies for U.S. capabilities and counter-capabilities in 2020, a further 10-year backward projection is made to identify the needed S&T activities that must be underway today to enable these.

In practice the precise dates must necessarily be interpreted somewhat notionally. In the cyber domain, for instance, it is essentially impossible to project technology 10 years into the future, and in that case the process was instead applied in a “5+5” framework.

Similarly, while this process has formed the foundation of the study, its implementation has been as much qualitative as quantitative. A fully quantitative assessment for each of the technologies considered here would have been beyond the scope even of this effort. Moreover, by ensuring that working groups and other sources of inputs to this effort included subject matter experts with technical and operational knowledge, quantitative aspects of this process could be addressed in part by assessments of these 10-year projections based on technical and operational judgment.

Identifying technology focus areas that will be most essential over this period has as much to do with technical and operational judgment as with direct quantitative analysis. As with any vision, its nature prevents it from being reducible to an algorithm that might seek to obtain similar insights. An articulation of the S&T that will be most valuable for the Air Force over the next two decades is ultimately the result of a set of reasoned, technically rooted, and objective judgments informed by an understanding of their strategic context.

3.3. Broad Range of Inputs to "Technology Horizons"

As Figure 3 shows, "Technology Horizons" consists of five phases, including three working phases designed to obtain a broad range of subject matter expertise in the implementation of the underlying process summarized above. Phases 2 and 3 included working groups, one each in the air, space, and cyber domains, and a fourth group that focused on cross-domain insights. The functions of these groups are described in Appendix B.1.1, their compositions are given in Appendix B.1.2, and their participant backgrounds are summarized in Appendix B.1.2. The working groups included representation from:

- Air Force S&T community
- Intelligence community
- MAJCOMs
- Product Centers
- FFRDCs
- Defense industry
- Academia

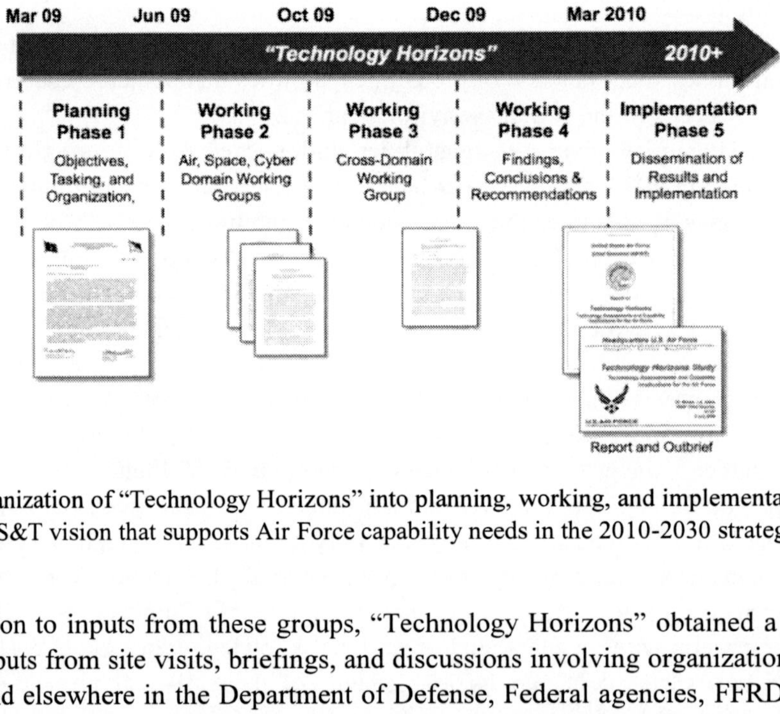

Figure 3. Organization of "Technology Horizons" into planning, working, and implementation phases to provide an S&T vision that supports Air Force capability needs in the 2010-2030 strategic environment.

In addition to inputs from these groups, "Technology Horizons" obtained a broad range of further inputs from site visits, briefings, and discussions involving organizations across the Air Force and elsewhere in the Department of Defense, Federal agencies, FFRDCs, national laboratories, and industry. These included inputs from Air Staff and Air Force Secretariat offices, MAJCOMs and Product Centers, Direct Reporting Units and Field Units. A list of these additional sources of inputs is given in Appendix B.2.

Additional operational perspectives were gained from briefings, visits and discussions with:

- Air Combat Command (ACC)
- Air Force Special Operations Command (AFSOC)

- Air Force Space Command (AFSPC)
- Air Mobility Command (AMC)

The effort further made use of a wide range of perspectives in nearly 200 reports and technical papers from the Air Force Scientific Advisory Board, the Defense Science Board, and other organizations. A partial list of these additional sources is given in Appendix B.3.

This exceedingly broad range of inputs to "Technology Horizons" has allowed development of a forward-looking, balanced, and effective S&T vision for the Air Force in 2010-2030.

3.4. A Focus on Cross-Domain Insights

While the working group phases were organized along the air, space, and cyber domains, it was recognized that some of the most important results from "Technology Horizons" would deal with cross-domain insights. A further working group was thus formed to address cross-domain effects, defined here to include:

- A technology in one domain that can produce unexpected effects in another domain; the effects could be beneficial or detrimental.
- A technology in one domain that requires supporting functions in another domain, and thus creates interdependences between two or more domains.
- A technology that falls "between" the classical domains but has implications in one or more of them.

Given the emphasis on cross-domain insights in this effort, the results from "Technology Horizons" are intentionally not presented along traditional domain boundaries.

The understanding and exploitation of cross-domain effects will grow in importance in the next decade and beyond, and the organization and execution of Air Force S&T will need to ensure access to such cross-domain insights and capabilities.

4. Organization of Results from "Technology Horizons"

The results from "Technology Horizons" are organized in the following three Volumes.

4.1. Volume 1: Vision for Air Force S&T During 2010-2030

This volume gives the elements of the S&T vision from "Technology Horizons" in Figure 4. These include (i) a set of overarching themes to guide Air Force S&T over the next decade, (ii) a set of key technology areas that will be among the most important S&T to emphasize during this time, (iii) a set of "grand challenge" problems to aid in focusing and integrating the research in these areas, and (iv) a plan for implementing the results from this effort.

Section 2: Strategic Context for Air Force S&T

Summarizes the principal strategic factors that will drive needed Air Force capabilities during this time, and factors relevant to the worldwide S&T arena that will impact the Air Force's ability to maintain superior technological capabilities over this time frame.

Section 3: Enduring Realities for the Air Force

Recognizes key drivers that will remain largely unchanged during this time and that will constrain the Air Force's ability to shape itself for the future using S&T in ways it might in an otherwise unconstrained environment.

Section 4: Overarching Themes for Air Force S&T

Identifies specific overarching themes that will be central for meeting Air Force needs dictated by the above strategic context and enduring realities; these themes form the essential foundation for the most important Air Force S&T during 2010-2030.

Section 5: Technology-Enabled Capabilities for the Air Force

Defines a set of technologically achievable capability areas that are well aligned with the overarching themes identified above, and then uses these notional capabilities to identify key enabling technologies that are most impactful across this set.

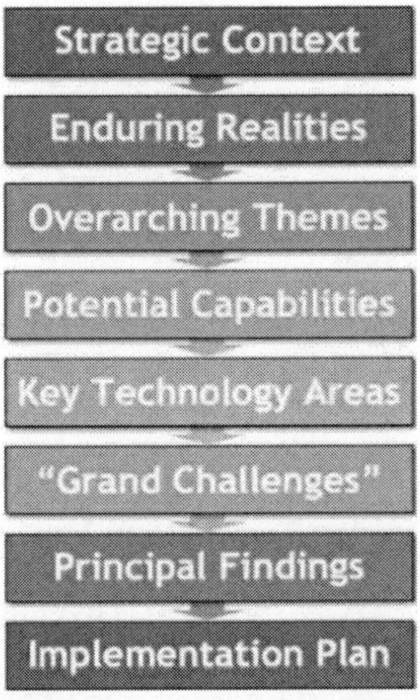

Figure 4. Major elements of the "Technology Horizons" vision for Air Force S&T over the next decade.

Section 6: Key Science & Technology Areas during 2010-2030

Determines a cross-cutting set of enabling technology areas that are most determinative over the notional capabilities identified above, and identifies them as being among the most important areas to emphasize in Air Force S&T over the next decade and beyond.

Section 7: Grand Challenges for Air Force S&T

Lists a set of challenge problems that will help focus Air Force S&T over the next decade on the key technology areas identified above through technology development efforts followed by systems-level integrated technology demonstrations to achieve stretch goals.

Section 8: "Technology Horizons": Principal Findings

Summarizes the major science and technology focus areas that the Air Force should emphasize during the next decade and beyond to enable technologically achievable capabilities that will give it the greatest Joint force effectiveness by 2030.

Section 9: Implementation Plan and Recommendations

Outlines the proposed plan for implementing the S&T vision from "Technology Horizons", provides actionable recommendations to vector Air Force S&T over the next decade, and identifies corresponding primary and supporting organizations to implement these.

> *The above sections in Volume 1 of "Technology Horizons" give the results of a rational approach for determining —from the enormous range of technologies that could benefit the Air Force — a set of guiding principles for the most important S&T to be conducted over the next decade.*

4.2. Volume 2: Current S&T in the Air Force Research Laboratory

Volume 2 summarizes how Air Force science and technology have been organized, planned, managed and conducted in the Air Force Research Laboratory over the past five years, as well as the core technical competences currently represented in AFRL and the overall body of research currently on-going in AFRL. An extensive appendix to Volume 2 gives detailed roadmaps of research efforts currently underway in AFRL.

4.3. Volume 3: Classified and Proprietary Supporting Material

Volume 3 provides certain SAP-level and proprietary supporting information. This material has been separated out to allow the widest possible distribution of Volumes 1 and 2.

5. Caveats

5.1. Value of Technologies Not Specified in "Technology Horizons"

No effort such as this can usefully list all S&T that is important for the Air Force to pursue, nor has it been an objective of "Technology Horizons" to do so. Instead, this effort has sought to identify some of the most valuable technology areas for Air Force S&T -- ones that the Air Force must pursue to enable key capabilities that it will need to be as effective as possible in the strategic environment during 2010-2030. The key technology areas identified herein thus necessarily represent only the most essential fraction of the overall research portfolio that should be pursued by the Air Force.

5.2. Focus on Key Technology Areas vs. Broad Technical Domain Areas

"Technology Horizons" was designed to identify key technology areas that are among the most essential for Air Force S&T to focus on over the next decade and beyond. These technology areas aggregate individual narrow research topics into usefully defined focus areas that Air Force S&T should emphasize, but avoid dictating the individual lines of research within these areas might prove to be most productive. Focusing on such key technology areas is the appropriate level of specificity for an S&T vision.

At the other extreme, overly broad domain area descriptors such as "material science", or even broad technical domains within these, such as "nanotechnologies", do not provide a comparable level of specificity needed to usefully serve as a guide for Air Force S&T. Key technology areas identified as most essential for the Air Force to pursue are sufficiently specific to usefully guide S&T investment choices, but are not so narrow as to dictate individual research projects to be pursued.

5.3. Connection between Air Force S&T and the Acquisition Process

Air Force S&T is the initial phase of the process by which technologies are matured and, where appropriate, are transitioned for acquisition by the Air Force. As a consequence, often-discussed improvements to the defense acquisition process could potentially improve the transition of technologies from S&T into systems for acquisition. At least some of the elements of the S&T vision identified in "Technology Horizons" could potentially achieve greater likelihood of practical impact from improvements to the defense acquisition process. However, attempts to suggest such improvements to the acquisition process were outside the scope of this effort.

SECTION 2. STRATEGIC CONTEXT FOR AIR FORCE S&T DURING 2010-2030

Air Force S&T can be prudently vectored only when placed in the context of the strategic environment in which the Air Force will be operating during this period. This Section summarizes key elements of this environment that will drive needed Air Force capabilities over the next two decades.

The strategic environment that the Air Force faces over the next two decades is substantially different from that which has dominated throughout most of its history. Since the end of the Cold War, long-standing paradigms that had ordered thinking and policy with respect to conflict have been replaced by a far broader range of threats and a less predictable set of challenges. Potential conflict scenarios now include not only major powers and regional players, but also failed and failing states, radical extremists, and a range of non-state actors that spans from organized militias, informal paramilitary organizations, warlords and warring ethnic groups to pirates, organized criminals, terrorists, and even individuals who may come from around the globe or across the street. Potential drivers of conflict range from religious extremism and ethnic disputes to resurging nationalism and aspirations for regional influence, and from competition over energy and natural resources to the need to contain nuclear proliferation and the spread of chemical, biological, and radiological weapons.

Unable to compete in direct engagements with U.S. Joint forces, adversaries have learned to exploit newfound asymmetric advantages. Some have sought to shift the battle into the media, exploiting public intolerance for real or contrived collateral damage -- strikes that by historical standards are being conducted with near surgical accuracy are now often deemed unacceptable. Others have learned to exploit U.S. dependence on the space and cyber domains, developing ways to disrupt or deny these to achieve potentially far-reaching cross-domain effects. Some are seeking access to nuclear capabilities. The Cold War, by comparison, seemed remarkably simpler, and the set of capabilities that the Air Force needs to be effective in the foreseeable strategic landscape is substantially different and appears less certain than ever before.

To aid in deterring conflict with major world powers, the U.S. must correctly understand their potential capabilities in the 2030 time frame, many of which are derived from investments in technologies that support development of military systems. China, for example, has focused much of its recent military modernization on investments in high-end asymmetric capabilities, emphasizing electronic warfare, cyber warfare, counter-space operations, ballistic and cruise missiles, advanced integrated air defense systems, and theater unmanned air vehicles. By combining imported technologies, reverse engineering, and indigenous development, it is seeking to rapidly narrow the technology and capability gap between the PLAAF and the U.S. Air Force. China is also using "military diplomacy" to expand security cooperation activities with Asian states and engage foreign militaries in a range of cooperative activities. Russia as well is developing and fielding new technology-derived military systems, including advanced fighter aircraft such as the T-50 PAK FA and advanced integrated air defense systems such as the SA-21. Global diffusion of advanced military systems through arms transfers, not only by Russia and China but also by France, Israel and many others, is making substantial technology-derived capabilities available around the world.

Among these, China in particular is widely regarded as having both the economic resources to devote to such advanced technology development programs and the national desire to achieve the resulting regional and global influence that such systems may bring. Its growing economy has helped pay for a massive military modernization program that includes not only new fighter aircraft, but also naval vessels and missiles.

Potential conflicts involving rogue nations, failed and failing states, radical extremists, and a range of non-state actors will likely remain one of the key drivers of U.S. Joint force needs.

Yet at the same time, China's one-child policy has created serious demographic imbalances in gender and age distributions, which are requiring it to direct resources toward social programs that could potentially constrain these ambitions. As a result, carefully chosen cost-imposing strategies developed through appropriate U.S. technology-enabled capabilities can provide an opportunity to slow the potential threat that this military build-up creates. By understanding the strategic environment facing the Air Force over the next two decades, the right choices can be made to obtain the mix of capabilities that will best meet U.S. national security objectives.

1. Relation to National Security Objectives

Military forces exist to develop and offer a range of operational and strategic options to the President for meeting national security objectives, and to Joint Force commanders for meeting military objectives. Throughout its history, it has been the consistent dual purpose of the U.S. military to not only fight and win the nation's wars but also to protect the nation and its global interests in ways that extend beyond direct combat operations. The latter acknowledges that military forces are as much an instrument employed for shaping the global environment, for deterring those who might otherwise harm the nation's interests, and for providing regional stability where needed to advance the nation's broader interests as they are for warfare itself.

At the broadest level these functions include:

- Protecting and defending the homeland from external attack.
- Deterring conflict with major global powers and encouraging regional stability through use of Air Force global vigilance, global reach, and global power.
- Preparing to fight and win the nation's wars as part of the Joint force.
- Providing military assistance to civilian authorities.
- Enabling national and partner instruments of power to prevent or contain local or regional instabilities.
- Supporting the nation's cooperative relationships with international partners and its interactions with competitors.

Supporting U.S. Joint force operations in so-called irregular or hybrid warfare while preparing for possible larger conflicts with a near-peer adversary is among the Air Force's greatest challenges.

High-end systems such as the F-22 support Air Force air dominance over the near- to mid-term; "Technology Horizons" seeks to define the S&T that will enable the next generation of Air Force capabilities that are suited to the needs of the 2010-2030 strategic environment.

In fulfilling these objectives to protect the nation and advance its global interests in times of peace, crisis, and combat, the Air Force is distinct among the Joint services in its ability to use the speed, range, flexibility, precision, and lethality of aerospace forces to provide global vigilance, global reach, and global power. Its mission requires it to look at the world from an inherently strategic perspective as it conducts global, regional, and tactical operations in the domains of air, space, and cyberspace in concert with the other Services, with other national instruments of power, and with U.S. international partners. This requires capabilities for the full spectrum of non-conflict and conflict operations, ranging from emergency response and humanitarian relief to counter-insurgency operations, major warfare and homeland defense.

2. Technology-Derived Challenges to Air Force Capabilities

The immense asymmetric advantage that the Air Force's air dominance has for much of the past 50 years provided for U.S. and partner forces could be potentially put at risk by worldwide development and proliferation of numerous advanced technology-derived threats, including integrated air defenses, long-range ballistic missiles, and advanced air combat capabilities. There have been equally important advances in counter-space technologies, in cyber warfare technologies, and in understanding the cross-domain effects that these technologies can produce on the U.S. ability to conduct effective air, space, and cyber domain operations. The combined effects of these and yet further technology advances that will occur over the next decade, and their transition into worldwide military systems, are essential elements of the strategic context that determines the most essential Air Force S&T over this period.

2.1. Advanced Electronic Systems

Land-based air defenses and air combat systems in particular are benefitting from increasingly widespread use of advanced electronics, including AESA radar technologies to enable more capable detection and targeting capabilities. These have evolved to counter non-traditional threats such as remote-piloted aircraft and cruise missiles. Advanced electronic warfare systems, including approaches based on digital radio frequency memory systems, that are being developed and proliferated today have shown significant abilities even against advanced AESA radars. Further advances in miniaturization and speed of electronics, and massive processing capabilities that these are enabling, are likely to yield further substantial increases in the capabilities of electronic warfare systems over the next decade and beyond.

Advanced fighter development efforts, such as the fifth-generation Russian/Indian T-50 PAK FA, could pose a significant mid- to long-term challenge to U.S. Joint air dominance.

2.2. Advanced Missile Seekers

Advances in electro-optical tracking, such as dual-band EO/IR imaging arrays with wide fields of view and advanced digital tracking filters, can be expected to significantly increase over the next decade. Early implementations of combined EO/IR imaging seekers in systems such as the highly maneuverable beyond-visual-range Python 5 air-to-air missile have shown the effectiveness of these technologies. Corresponding implementations of these advanced technologies in surface-to-air missiles, such as the SPYDER anti-aircraft missile system based on the Python 5, are extending these technology advances to area defense systems. Such lethal systems are already beginning to be proliferated via international sales. Their capabilities will appear even more broadly over time, both through indigenous developments that use these technologies and through subsequent global sales of such systems.

2.3. High-Speed Airbreathing Missile Propulsion

High-speed airbreathing missile propulsion technologies have steadily advanced over the past two decades and are now enabling substantially more capable supersonic air-to-air missiles, surface-to-air missiles, and cruise missiles. Such a ramjet-derived propulsion system is used in the MBDA Meteor, a Mach 4+ air-to-air missile that is likely to enter service in 2013. A ramjet derivative propulsion system is also being used in the Russian/Indian BrahMos Mach 2.5+ cruise missile, brought into service in 2007, and a related ramjet-based airbreathing propulsion system is used in the Indian Mach 2.8+ Akash surface-to-air missile first fielded in 2009. The speed, range, and terminal maneuverability benefits of such airbreathing supersonic propulsion technologies will likely accelerate development of missile systems to exploit these advantages.

International sales of advanced IADS, such as the SA-21A shown at right with 92N2E engagement radar and 96L6E surveillance radar, could present potential anti-access/ area-denial capabilities.

2.4. Advanced Integrated Air Defense Systems

Development and proliferation of advanced integrated air defense systems have progressed substantially. Longer-range mobile surface-to-air missiles in the Russian S-300 family, such as SA-10s/20s, which have been exported widely, and corresponding Chinese manufactured derivatives of "double digit SAMS" such as the HQ-10/15/18, represent significant challenges. These systems can potentially engage legacy fighters at ranges beyond their own ability to hold such targets at risk. More advanced Russian systems such as the S-400 have advanced tracking and longer-range capabilities, and could potentially be sold internationally. Modern integrated systems such as these are enabling a fundamental change in air defense strategy, from traditional point defense of key targets to broader anti-access/area-denial approaches based on offensive and defensive counterair operations.

2.5. Passive Sensors and Electronic Warfare

Further development of passive sensor technologies such as infrared search and track (IRST) systems over the next decade or more will make air defense systems increasingly resistant to electronic suppression. Potential adversaries are also upgrading to networked air defense systems with advanced electronics and signal processing capabilities that can make jamming far more difficult. AESA radars are allowing active electronic beam forming and steering. Electronic warfare capabilities of potential adversaries are increasingly shifting from jamming to sophisticated spoofing, made possible largely by the continuing miniaturization of commercial electronics. Availability of inexpensive yet massively capable electronic processing and storage allows the number and sophistication of such devices to grow rapidly, and to be fielded in relatively low-cost EW pods and decoys. Software changes allow these threat systems to be readily adapted to new countermeasures.

Technologies such as the ramjet-derivative propulsion system on the MDNA Meteor are beginning to enable high-speed air-to-air missiles with greater range and terminal maneuverability.

At the same time, there has developed and overlap between electronic warfare and cyber warfare, especially in the wireless RF domain. The technological sophistication, diversity, and proliferation of advanced electronic warfare systems present a substantial challenge to U.S. air dominance. New technology-derived approaches will be needed across the spectrum of standoff and close-in EW capabilities, including electronic attack, protection, and support techniques suited to permissive, contested and highly-contested environments.

2.6. Shoulder-Fired Surface-to-Air Missiles

Next-generation shoulder-launched surface-to-air missile systems can be expected to further add to the threats that advanced technologies pose. Current "low end" man-portable air defense systems (MANPADS) are affordable enough to buy in large quantities and can be readily dispersed, hidden, and employed for close-in air defense. Further advances in miniaturization of electronics and processing are likely to produce substantially more capable systems at price points that may permit comparably widespread global diffusion.

Dual-mode EO/IR focal-plane array seeker technology, as on the Python-5 air-to-air missile, can give wide field-of-view imaging infrared capability and full-sphere beyond-visual range engagement.

2.7. Ballistic and Cruise Missiles

The emergent long-range air defense capabilities of several potential adversaries have serious implications for U.S. airpower. Additionally, ballistic and cruise missiles with growing mobility, range, maneuverability, and precision pose further threats to Air Force systems and their ability to deliver the power projection on which much of U.S. strategy is based. This is especially the case in the Western Pacific, where China has fielded conventional ballistic missiles and cruise missiles with significant reach. By 2015 China is expected to have hundreds of DF-15 ballistic missiles and DH-10 cruise missiles capable of reaching much of the western Pacific. Advanced medium-range ballistic missiles such as the DF-21, DF-25, and the intercontinental-range DF-31 may be in service in significant numbers, and a DF-41 advanced intercontinental ballistic missile may be under development.

The threats posed to U.S. air bases in the Pacific by ballistic missiles create significant technology challenges for maintaining air power projection in this region. The long ranges required for operations in the Pacific entail substantial tanking requirements that put a premium on long-range strike capabilities and fuel-efficient propulsion systems. The air-sea battle concept based on synchronized Air Force and Navy operations against a potential near-peer competitor in this region may also require significant new capabilities.

2.8. Advanced Fighter Aircraft

Russian development of the Sukhoi T-50 PAK FA fifth-generation fighter, reportedly in a joint effort with India, is an indicator that the U.S. must continue to develop its capabilities to maintain air dominance. The aircraft will reputedly include substantial technology-derived advances over all prior potential competitor aircraft, including significant low-observable capability, advanced AESA radar, advanced avionics, and supercruise capability. Early flights of the PAK FA in January and February 2010 suggest a serious development program that has reached an advanced stage of engineering refinement. Reported Chinese efforts to develop one or more J-XX fifth-generation fighters appear not to be as far advanced, but the resources that may be available to devote to these efforts suggest such a capability could be fielded within the time horizon of this study. In general fifth-generation fighter systems, if deployed against the U.S., have the potential to pose substantial challenges to our air dominance over the next decade.

Advanced surface-to-air missile systems, such as the SPYDER that combines Python-5 EO/IR missiles with active radar-guided missiles, can provide highly capable area defense capabilities.

2.9. Remote-Piloted Aircraft

Growth in military use of remotely piloted vehicles has been rapid as forces around the world explore increasingly wider uses for them, including surveillance, strike, electronic warfare, and others. These will include fixed-wing and rotary-wing systems, airships, hybrid aircraft, and other approaches. They will have increasingly autonomous capabilities allowing remote pilots to declare their overall mission intent but permit these systems to adapt autonomously in the local environment to best meet those objectives. Some systems may operate collaboratively in multiple-craft missions to increase survivability and deliver effects that could not be achieved individually. Technologies are enabling such increasingly autonomous systems to exchange data on their respective states in order to adjust their mission planning and adapt to their changing environment. Autonomous aerial refueling capability will enable long-range and long-endurance operations. While most remote-piloted systems will likely have limited capabilities, some will incorporate low-observable technologies, advanced electronic warfare functions, and directed-energy strike capabilities. Price points of some systems could allow them to be acquired in numbers sufficient to present significant challenges for certain types of missions.

The stand-up of Air Force Global Strike Command (AFGSC) has consolidated all Air Force assets for the nuclear mission, including B-52s and B-2s, under a single major command.

2.10. Directed Energy Systems

Laser-based and high power microwave-based directed energy systems being developed by several nations will play an important role in the strategic context of 2010-2030. MDA's success in tracking, targeting, and destruction of a representative missile in February 2010 with an integrated, airborne, megawatt-class chemical laser demonstrated the potential of strategic-scale systems. Ground-based lasers are likely to appear for air defense and other roles, as will airborne microwave-based systems that can disable or defeat electronic systems. More recent solid-state laser technologies are enabling tactical-scale systems for potentially revolutionary airborne self-defense and low collateral damage strike capabilities. Emerging fiber laser technologies as well as diode-pumped alkali lasers may allow later versions of such systems to be made even smaller for integration in a much wider set of platforms, including fighters. Directed energy systems will be among the key "game-changing" technology-enabled capabilities that enter service during this time frame.

2.11. Space Control

U.S. Joint force dependence on the space domain is a further key component of the strategic context for Air Force S&T during 2010-2030. Reliance on assured access to military space systems is essential to meet the demand for global intelligence, surveillance, reconnaissance, and communications. Potential adversaries recognize our critical dependence on space assets and are developing means for disrupting our access to those assets, or even for incapacitating or destroying those assets. Jammers allow interference with satellite uplinks and downlinks, and cyber attacks can disrupt ground control segments. Some are developing the means to gain access to space themselves. Others are making use of increasingly capable commercial space-based imaging capabilities, and space-based imagery freely available via the Internet at levels of resolution that not long ago were accessible only to the intelligence community.

Continuing to strengthen the Air Force nuclear enterprise remains the top U.S. Air Force priority; here Airmen load AGM-129A advanced nuclear cruise missiles onto a B-52 at Minot AFB.

2.12. GPS Denial

The broad dependence of U.S. Joint warfighting capabilities on precision navigation and timing from GPS satellites has made local or regional GPS denial a high priority for adversaries. Beyond preventing access to positioning information, inexpensive low-power GPS jammers and spoofers can deny accurate timing information that is at least equally critical to U.S. systems. China's efforts to field its Beidou-2 (Compass) PNT satellite network and Russia's fielding of its GLONASS system suggest the role that navigation warfare via GPS denial and spoofing may play in the strategic thinking of these and other nations.

2.13. Space Launch Capabilities

Altogether there are 11 nations today that have orbital launch capability, and roughly 50 that possess orbiting satellites. In 2009 with the launch of its Omid telecommunications satellite, Iran became the ninth country to have developed the capability to place indigenously developed satellites into low Earth orbit using indigenously developed launch systems. This was achieved with its Safir-2 space launch vehicle, a derivative of the Iranian-developed Shahab-3 medium-range ballistic missile that itself is believed to trace to the North Korean Nodong-1 MRBM. A subsequent Iranian launch in February 2010 showed the capability of its Kavoshgar-3 rocket.

Ballistic missiles such as the Chinese medium-range DF-25 (left) and intercontinental DF-31A (right) demonstrate significant capabilities in terms of range, mobility, and accuracy.

Iran further claims to be developing a larger Simorgh orbital launcher, and North Korea too is seeking to develop a satellite launch capability. Countries with these sophisticated capabilities could potentially develop intercontinental ballistic missile systems. Moreover, private corporations now operate 19 launch sites throughout the world, providing launch options for nations seeking to gain access to space. In 2009, the first orbital launch using a commercially developed liquid-fueled rocket occurred when Space-X put a Malaysian satellite into low Earth orbit using its Falcon-1 launcher, and its EELV-class Falcon-9 rocket had its first launch in 2010.

2.14. Satellite Technologies

Technologies are also enabling a range of 'small sats' with masses below about 200 kg, some of which can already today provide significant military capabilities. They include microsatellites with masses between 10-100 kg, nanosatellites with masses between 1-10 kg, and picosatellites with masses below 1 kg. The latter include "CubeSats" measuring just 10-cm on a side, which meet published design standards that allow their deployment by common low-cost mechanisms. Combined materiel and launch costs for such systems are as low as $100K. They are drawing extensive interest from universities, companies, and others around the world, and numerous such systems have been launched. Efforts are underway to develop low-cost, standardized, on-demand orbital imaging systems for such small satellites. They represent a further aspect of the increasing low-cost access to space that is available to numerous potential adversaries.

2.15. Direct-Ascent ASAT Capabilities

Offensive counterspace capabilities are also growing, most notably in anti-satellite warfare capabilities. In 2007 China became the third known country with a proven ASAT capability when it conducted an unannounced launch of a modified DF-21 intermediate-range ballistic missile designated the KT-1 space-launch vehicle to destroy its own defunct Feng Yun-1C meteorology satellite. Essentially any nation with a space launch capability could potentially field a rudimentary ASAT program. The dual use of civilian and military rockets being developed by several countries, including Iran, North Korea, India and Israel, could lead to rapid growth in the number of players with the technical capability to develop ASAT systems.

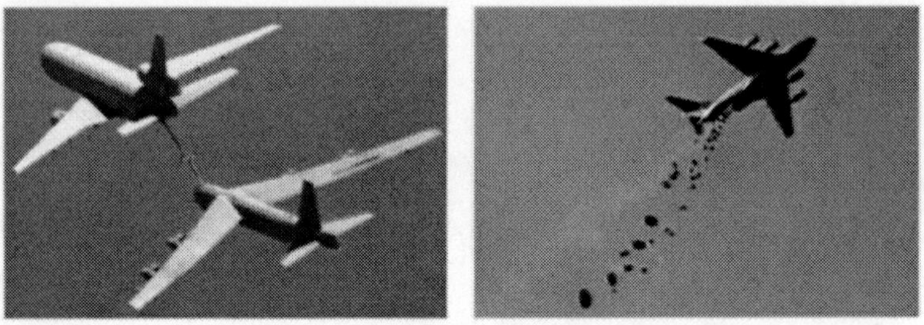

Air mobility operations including aerial refueling and airdrop will remain both challenging and increasingly critical for supporting Air Force and U.S. Joint force operations.

2.16. Directed-Energy ASAT Technologies

China's demonstrated direct-ascent ASAT capability comes in addition to open-source reports of Chinese and Russian efforts to develop directed-energy anti-satellite capabilities, primarily in the form of ground-based lasers. Even relatively low-power lasers can temporarily blind or dazzle electro-optical and infrared sensors in ISR satellites. At somewhat higher power levels they can permanently disable these detector arrays, or can degrade the effectiveness of solar panels and other components to render a satellite inoperable. At sufficiently high power they can induce thermal overloads in satellites or even destroy them directly. The rapid time scales on which these effects can be achieved make satellite self-protection extremely challenging. In principle such ground-based laser ASAT systems can be made mobile and widely dispersed, further complicating satellite protection at the terrestrial end.

2.17. Co-Orbital ASAT Systems

Beyond direct-ascent kinetic-kill ASAT capabilities and ground-based lasers for directed-energy ASAT capability, co-orbital satellites represent a further emerging ASAT concern. Such systems might be relatively small satellites designed to provide an on-demand kinetic kill capability, or might have laser-based or microwave-based directed-energy capabilities to degrade or destroy satellites. Co-orbiting satellites could also provide an adversary counterspace options beyond ASAT capability, for instance by interfering at relatively close ranges with satellite uplink and downlink transmissions. Such small, maneuvering, co-orbiting satellites also offer many other options for lethal and non-lethal "proximity operations" in support of counterspace efforts.

2.18. HAND Effects on Space Systems

Beyond the well-known prompt EMP effect of a high-altitude nuclear detonation that affects terrestrial and airborne electronics, there is a further long-term effect that can specifically disable space systems. Such a nuclear detonation would act to populate Earth's Van Allen radiation belts with large numbers of energetic electrons produced from beta decay of fission fragments. These high-energy electrons would remain trapped for years by Earth's magnetic field. Satellites in LEO or HEO orbits would be disabled from effects of the ionizing electrons on critical satellite parts over months or years as they pass through the resulting enhanced radiation belts. This could occur as a side effect of a regional nuclear war or as a deliberate effort by a rogue nuclear adversary seeking to cause massive economic damage to the industrial world.

Increasingly capable remote-piloted aircraft systems are being used worldwide, while directed energy systems may achieve tracking, targeting, and destruction of such systems.

2.19. Orbital Debris

Increasing amounts of orbital debris also pose serious threats to space assets, as demonstrated by the February 2009 collision between a functioning Iridium-33 satellite and a defunct Russian Kosmos-2251 satellite. The collision itself produced nearly 400 new pieces of orbital debris. By comparison the 2007 Chinese kinetic-kill ASAT demonstration produced between 20,000 and 40,000 debris pieces. There are in fact some 1600 conjunctions predicted every day within the accuracy of current analyses. Just two weeks prior to the Iridium-Kosmos collision a discarded 4-ton Chinese upper-stage rocket body passed within 48 meters of the 8-ton European Envisat Earth-observation satellite, producing a 1-in-70 chance of a collision that by itself would have doubled the entire catalog of human-made objects in orbit.

2.20. Cross-Domain Vulnerabilities of Air, Space and Cyber Systems

Space has long been recognized as an operational domain with capabilities far beyond the original supporting functions that it provided for the air domain. The cyber domain as well has come to be recognized as a domain of its own beyond the supporting functions that it originally provided for the air and space domains. Yet Air Force systems and operations in the air, space and cyber domains are highly interdependent. Resulting cross-domain effects associated with this interdependence have only recently begun to be understood. They include opportunities for gaining synergies through properly orchestrated combined operations in all three domains to achieve more than the sum of individual operations. However they also include far-reaching cross-domain threats that arise from these interdependences, many of which are only recently being widely recognized. Potential adversaries will increasingly seek to exploit these cross-domain influences. The challenge is to continue gaining the synergistic cross-domain benefits of these interdependences while minimizing the potential vulnerabilities that they can create.

A growing number of countries are gaining access to space using indigenously developed launch systems, such as the Iranian Simorgh, or through commercial launch services worldwide.

2.21. Cyber Domain Challenges

While the cyber domain was originally envisioned mainly in the context of computer software, hardware, and networks, it is now recognized as encompassing the entire system that couples information flow and decision processes across the air and space domains. It thus comprises traditional wired and fiber-optic computer networks based on electromagnetic waveguides, but also free-space wireless transmissions of voice, data and video at radio frequencies, optical wavelengths, and other parts of the electromagnetic spectrum, in addition to the hardware and software associated with such systems. Viewed in this way, the enormous cross-domain opportunities and threats presented by the cyber domain become more clearly apparent.

2.22. Coupling of Cyber, EW, and ISR Vulnerabilities

Cyber domain opportunities and vulnerabilities apply not only to traditional network warfare and information warfare, but extend to all aspects of electromagnetic spectrum warfare. The latter encompasses certain aspects of electronic warfare, some of which in turn have close synergies with intelligence, surveillance and reconnaissance and other operations in both the air and space domains. This strong coupling of all three Air Force operational domains through the electromagnetic spectrum reveals how essential freedom of operation and control of the spectrum will be both for protecting against cross-domain vulnerabilities and enabling the potentially enormous benefits that can come from cross-domain synergies.

The disproportionate U.S. dependence on cyber-enabled systems allows potential adversaries to gain an asymmetric advantage at comparatively low cost via development of offensive cyber capabilities.

Potential adversaries have recognized these cross-domain vulnerabilities and the opportunities they could present to them. Because U.S. Joint force operations have evolved to become highly dependent on the cyber domain, the strong integration and interdependence of all three Air Force operational domains provide most potential adversaries with an inherently asymmetric advantage. They have found that attacks through cyberspace allow them to create tactical, operational, and strategic effects at low cost and with relative impunity. Most see the cyber domain as a relatively inexpensive venue within which they can potentially offset many of the Air Force's current overwhelming advantages in the air and space domains.

For instance, as China undertakes its force modernization, it does so at a time when the cyber threats and opportunities are being understood, and thus has an asymmetric advantage in being able to build a military cyber infrastructure that could potentially be less vulnerable than that which had been put in place much earlier by U.S. joint forces. Developing the ability to make Air Force operations resilient to cyber attacks -- as distinct from the current emphasis on protecting adversary access to cyber systems -- will become increasingly important.

2.23. Cyber Operations in Untrusted Environments

Other countries are increasingly becoming the majority providers of equipment and infrastructure for cyber systems. Particularly in cyber networking technologies, markets have judged foreign-based sources to be comparably capable and less costly than U.S. providers. This situation is causing critical mission traffic to pass over enterprise architectures that involve significant foreign control, and that therefore cannot be fully verified with regard to trust/integrity. This includes foreign suppliers of both hardware and software, creating risks to critical missions that must operate in inherently untrusted cyber environments. Although these cyberspace systems can greatly enhance mission effectiveness, a clear understanding of the resulting vulnerabilities they create in key mission areas is typically lost in the rush to gain the benefits cyber-enabled systems. As Air Force capabilities across the air, space, and cyber domains become increasingly reliant on such systems, a key challenge will be to effectively manage the risks associated with having to operate in environments that are inherently untrusted.

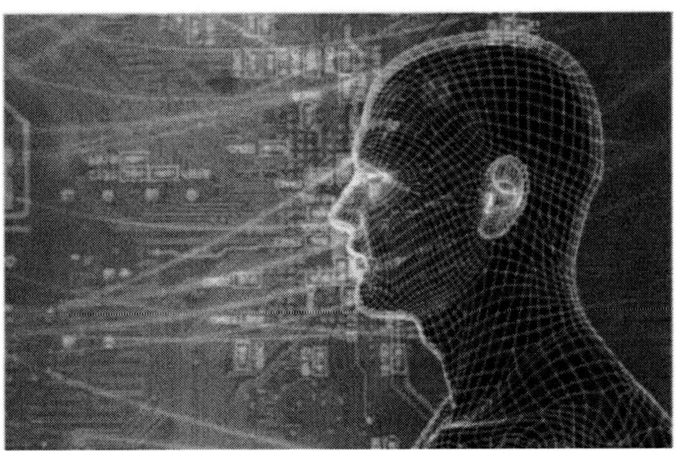

Technologies such as massive virtualization and rapid network recomposability that enable inherent resilience in cyber systems will be key to gaining the cost and capability advantages of autonomy.

3. Strategic Implications of S&T Globalization

Further aspects of the strategic context affecting Air Force S&T during this time originate from changes that are occurring due to globalization of science and technology. Over the next two decades, the U.S. will not dominate science and technology -- or the world economy -- the way it has over the past 50 years. Already today, the substantial S&T gap that once existed between the U.S. and much of the rest of the world is closing rapidly. Some say this is the result of U.S. inability to continue educating and advancing a competitive workforce in science, technology, engineering, and mathematics (STEM) fields. Yet regardless of how well the U.S. addresses that challenge, it is at the other end of the "gap" that most of the change is occurring.

The rest of the world has recognized science and technology as central to national economies that can compete effectively in the knowledge-based global marketplace. They have sent their best and brightest to be educated, often at U.S. universities, and many have returned home to build education systems that, over time, have allowed development of substantial indigenous science and engineering workforces. Some of these countries have also used internal resources to develop the industrial capacity needed to translate science and engineering knowledge into technology-derived products and capabilities. The effect of these investments is that the worldwide S&T gap will continue to close regardless of what the U.S. may do. The U.S. and Department of Defense will be competing on an S&T playing field that is substantially different than it was a decade or two ago.

In parallel with this, knowledge and information have become readily available worldwide. Just a decade ago, gaining access to the technical papers needed to understand the results of prior development efforts on any given topic around the world, and thereby identify the most productive avenues for further work, was a significant challenge and a substantial barrier to entry in most technical fields. Today the Internet puts literally thousands of technical papers published every day at nearly anyone's fingertips. The literature search needed to understand the status of any given field and determine the best ways to advance it may be only a few mouse clicks away. The net effect is a rapid and massive global diffusion of science and engineering knowledge that has fundamentally changed how technology advancements made on one side of the world can be translated into militarily-relevant systems on the other.

At the same time there has been a growing convergence of the key technologies involved in worldwide consumer products with the technologies needed for enabling many advanced military systems. This trend has been driven primarily by the role that advanced electronics technologies -- once largely the domain of high-end military hardware -- have come to play in the consumer products market. Today, the technical level of the electronics, processing, and software in many ordinary consumer devices may differ only marginally from those in many high-end military systems, and in some cases exceeds these.

The effect has been to create a global science and engineering workforce having technical skills that can be readily converted from developing consumer products to development of substantial military capabilities. Similarly, enormous investments being made worldwide in legitimate life sciences and biotechnology research are providing the knowledge and workforce that can enable others to develop closely related biological weapons capabilities in a relatively short time. In short, the advancing globalization of S&T is bringing knowledge, technology, and industrial capacity developed for the consumer market within reach of a far wider range of players, who can readily transform these into militarily capable systems. Others will find easy access to the resulting technology-derived military systems in international arms markets.

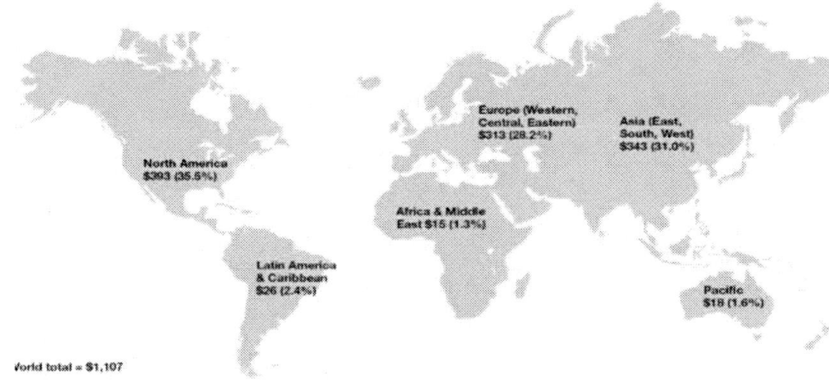

Regional R&D expenditures in $B and percentage of world total, showing North America, Europe, and Asia as roughly equal in R&D investment. (Adapted from 2010 NSF Science & Engineering Indicators.)

A further effect of the globalization of S&T and the resulting rapid pace at which technical advances are being made is that the lifetime of many such advances is becoming remarkably shorter. In part due to the sheer number of scientists and engineers now working worldwide in technical fields, the pace of both technical innovation and obsolescence has increased sharply. In the consumer products market, it is the innovation cycle that receives the most attention, but in the military systems market it may be the obsolescence cycle that is more important. As a result, the lifetime of at least some technology-derived military systems may not be nearly as long as it has been in past decades. This can be expected to impact defense acquisition choices as well as the S&T investment decisions needed to support these.

4. Federal Budget Implications for Air Force S&T Strategy

A final essential aspect of the strategic context that will guide Air Force S&T during 2010-2030 is the influence that Federal budget pressures will have. Combined impacts of the Federal debt, the aging population, and Federal entitlement promises made through Social Security, Medicare and Medicaid have begun what will over the next several decades become an unprecedented drain on the Federal budget that will increase rapidly in coming years. The first of 77 million retiring "baby boomers" born between 1946 and 1964 became eligible for Social Security benefits in January 2008. Their numbers will grow at a rate of 4 million per year through 2026, and they will continue to draw entitlement benefits through 2050.

The Congressional Budget Office has calculated that the cost of these benefits will grow from 8.4% of GDP today to 14.5% by 2030, and 18.6% by 2050. By comparison, the entire Federal budget today is just 20% of GDP. At current rates, CBO has estimated that by 2049 these entitlements would consume every other Federal program except interest on the Federal debt. Even with potential changes in entitlement benefits, the looming Federal budget pressures will be immense, and will necessarily impact the range of S&T and other Air Force efforts that can be pursued.

SECTION 3. ENDURING REALITIES FOR THE AIR FORCE DURING 2010-2030

This Section notes several enduring factors that will continue to influence Air Force S&T needs and the environment in which Air Force S&T will be conducted over the next two decades.

Beyond elements of the strategic environment in the previous Section, there are other key factors that represent enduring needs of the Air Force in 2010-2030, which form essential background for an S&T vision to guide investment choices over this period.

1. Ensuring Interoperability with Legacy Systems

In 2030, many manned and remote-piloted aircraft and other systems that the Air Force operates today will continue to be needed and remain in use. Ensuring interoperability with these systems, especially as the technologies in newer systems advance rapidly, will become an increasing challenge. Technologies that can be cost-effectively integrated in these systems to upgrade their capabilities, or that otherwise enable these legacy systems to interoperate effectively with newer systems, will become even more important than they are today.

2. Sustainment Costs for Legacy Systems

Sustainment is essential to the Air Force mission. As legacy air platforms and other systems continue to be used throughout this period, and as new platforms and systems are introduced during this time, technologies to support improved sustainability or to reduce costs associated with sustainment will continue to be essential.

3. Importance of Low-Observable Systems

Low-observable technologies and the systems that employ them for long range penetrating and persistent strike are among the most distinguishing elements of the Air Force. While advanced IADS may over time create an increasingly challenging environment for these critical systems, low observable systems will remain essential for the ability they give to penetrate defended airspace, for the sensitivities they demand in the air-defense systems of potential adversaries, and for the potential secondary benefits that this can create for other technology-based capabilities. Technologies to extend affordable LO capabilities will remain essential.

4. Energy Costs and Availability

Reducing Air Force energy costs, especially the costs associated with aviation fuel usage, and increasing the assured availability of energy, including fuels and base energy, will continue to be growing challenges during 2010-2030. As with other Air Force costs, fuel and energy costs will reduce the availability of resources that could otherwise be used to provide increased capability dominance. The need for advances in a broad range of technology areas to reduce fuel use, including propulsion, structures, aerodynamics, and controls, as well as technologies to allow increasing use of alternative fuels, will remain important, as will technologies to enable air base operations with increasingly greater independence from the electricity grid.

5. Growing Role of the Cyber Domain

During the past decade, no other technology domain has grown in importance as rapidly as the cyber domain. Understanding of potential vulnerabilities that cyber systems create has grown from defending computer networks to include network-enabled RF systems and other parts of the electromagnetic domain, as well as hardware and software supply-chain vulnerabilities. This domain has become centrally important for maintaining Air Force mission assurance and for enabling entirely new mission capabilities. While the defensive aspects of cyber warfare may improve somewhat as systems developed in the "pre-cyber" age are retired, cyber warfare at various levels will continue to grow rapidly in importance in the 2010-2030 environment.

6. Need for "Soft Power" Capabilities

Air Force missions have always included many types of other-than-conflict operations, including emergency response, humanitarian relief, peacekeeping missions, and contingency operations as well as joint exercises and other cooperative activities with international partners. The role of these missions has grown in recent decades, and will likely continue to expand over the next two decades. There will be an enduring need for technologies that can enable a wide range of Air Force capabilities that support these types of missions, including better multi-level security solutions to facilitate sharing of information, systems, and training with international partners.

7. Manpower Costs

Manpower costs associated with pay and benefits are consuming a significantly growing fraction of the total Air Force budget, pressuring resources for all other functions that support Air Force missions. Technologies that can enable the same or greater mission effectiveness with reduced manpower requirements can free up resources to permit greater overall Air Force capability. For example, these may include technologies for increasingly autonomous systems and methods to increase trust in them, for augmentation of human performance, for improved processing, exploitation, and dissemination of intelligence, surveillance and reconnaissance data, and for reducing maintenance and sustainment costs. The need to focus S&T efforts on these and other technologies to reduce manpower costs will continue to become increasingly important.

8. Budget Constraints

Air Force budgets throughout this period will likely remain constrained. As noted in Section 2, combined impacts of the Federal debt, the aging population, and Federal entitlement promises made through Social Security, Medicare and Medicaid have begun an unprecedented drain on the Federal budget that by 2030 will consume nearly three-quarters of the entire budget. Even if reductions in entitlement promises are enacted and Federal taxes as a percentage of GDP are raised substantially over today's rates, the effect of these factors will be to place increasingly severe limits on Air Force budgets over this period.

9. Effective National S&T Partnerships

The Air Force has consistently maintained working partnerships that include coordination of science and technology programs as well as collaborations on joint S&T efforts with its sister Services, DARPA, NASA, IARPA, and other organizations. The importance of such effective partnerships will continue and is likely to grow as budget pressures necessitate even greater efficiencies, and as the scale and scope of many ambitious efforts exceed the reach of any one organization alone. These benefits of such effective partnerships extend as well to FFRDCs and industry involved in S&T-related efforts.

10. Continued International S&T Cooperation

International collaboration in S&T has grown significantly over the past decade, providing access to capabilities that complement those domestically available. Building and expanding international S&T-related partnerships has been achieved, for example, through efforts such as the Engineers and Scientists Exchange Program and the "Windows on Science" program, through AFOSR's European (EOARD), Asian (AOARD), and Southern (SOARD) Offices of Aerospace Research and Development, and through SAF/IA's air armaments cooperation program. In addition to leveraging the limited resources available for advancing the U.S. technological base, these international cooperations help build partner capacity and can help in efforts to avoid capability surprise. Such international S&T cooperation programs will play an increasingly important role in the 2010-2030 environment.

11. STEM Workforce

Air Force capabilities are dependent on a workforce that has the needed science, technology, engineering and mathematics (STEM) backgrounds to provide the S&T advances on which these capabilities will be based. Development and maintenance of such a workforce will be essential for achieving the S&T vision laid out in "Technology Horizons". This applies to the workforce needed within AFRL as well as within the broader industrial base. The Air Force already makes important contributions to successful STEM programs focused at the K-12 and college/university levels through open-house efforts in AFRL and via direct mentoring relationships. Support of the Air Force's "Bright Horizons" plan for a coordinated STEM workforce development effort will need to be sustained to provide the workforce needed over the next decade and beyond.

SECTION 4. OVERARCHING THEMES FOR AIR FORCE S&T DURING 2010-2030

This Section identifies a number of broad themes that will serve to guide Air Force S&T in ways that can accelerate the delivery of technology-derived capabilities able to respond to the strategic environment, budget environment, and technology environment during this period.

The strategic and budget environments outlined in earlier Sections, together with the state to which key technologies can be brought during the next decade to meet the demands of those environments, lead to a set of broad technology themes that will be of overarching importance during this time. These themes provide guidance and direction for identifying S&T efforts that will be among the most essential in terms of their impact on Air Force capabilities.

These overarching themes are presented in Figure 5. Each row represents a significant shift in research emphasis within a given area, namely a decrease in emphasis on one area toward an increased emphasis in a different focus area. Both the reduced emphases in the left column and the increased emphases in the right column combine to ensure an S&T portfolio that is well suited to enabling key capabilities the Air Force will need during 2010-2030.

The trends in Figure 5 are fundamental shifts in research focus areas to guide Air Force S&T in directions that will provide it with the technologies needed to enable capabilities that are matched to the environment over the next decade and beyond. The shifts in research balance identified in Figure 5 should not be interpreted as a call for ending all research in the technical areas identified on the left, since this would create a substantially unbalanced S&T portfolio. Instead, substantially greater emphasis must be given to those on the right.

Beyond the impact on S&T efforts, the shifting emphases in Figure 5 point to a need for systems engineers in Product Centers to become familiar with the resulting technologies and to develop approaches and tools for integrating them in Air Force systems. Design protocols, subsystem interactions, and other aspects of system development all will be impacted by these changing characteristics of future Air Force systems.

1. From … Platforms to … Capabilities

Since the end of the Cold War and its single major nation-state threat scenario, emphasis has been increasingly shifting from "platforms" exquisitely designed to overcome a single threat toward "capabilities" able to meet a far wider range of threats. For instance, potential adversary anti-access/area-denial capabilities and long-range air-to-air missiles are increasing emphasis on standoff strike and beyond-visual-range engagements that rely less on platform-specific functions such as thrust vectoring for combat maneuverability and more on capabilities derived from the platform contents, such as ISR and electronic warfare functions. This trend is likely to continue, and it suggests that core technologies associated with platforms, such as aerodynamics and flight control, will continue to be needed but that greater emphasis will shift toward other technology areas that deal more directly with capability-providing functions.

Although platforms will remain significant drivers of capability, the challenges the Air Force faces require S&T to place increasing emphasis on capability-based efforts over platform-centric efforts.

2. From … Manned to … Remote-Piloted

The growth in Air Force and other DoD use of remotely piloted vehicles has been rapid, as their utility for operations far beyond surveillance has come to be understood. Such systems are also being used for an increasingly wider range of non-airborne applications. Although some types of emerging systems will still benefit from having one or more crew members on board, for many applications the weight, volume and endurance limitations associated with such crew, as well as the cockpit environmental control systems needed to support crew operations, extract an unacceptable performance penalty. Manned systems also present risk to crew that can be avoided in remote-piloted systems.

For these reasons, an increasing shift to remote-piloted systems is inevitable over the next decade, and technologies will be needed to permit the potentially greater functionality of such systems to be exploited for entirely new missions. Such systems are likely to involve varying degrees of autonomy, require endurance far beyond what is possible today, and include large high-altitude airships with massive ISR capabilities, all requiring new supporting technologies.

Technologies to enable an increasingly wider array of remote-piloted systems, many of which will bear little resemblance to today's systems, will become increasingly important over the next decade and beyond.

From decreased emphasis on this ...	To increased emphasis on this ...
1. *From* ... **Platforms**	*To* ... **Capabilities**
2. *From* ... **Manned**	*To* ... **Remote-piloted**
3. *From* ... **Fixed**	*To* ... **Agile**
4. *From* ... **Control**	*To* ... **Autonomy**
5. *From* ... **Integrated**	*To* ... **Fractionated**
6. *From* ... **Preplanned**	*To* ... **Composable**
7. *From* ... **Single-domain**	*To* ... **Cross-domain**
8. *From* ... **Permissive**	*To* ... **Contested**
9. *From* ... **Sensor**	*To* ... **Information**
10. *From* ... **Strike**	*To* ... **Dissuasion**
11. *From* ... **Cyber defense**	*To* ... **Cyber resilience**
12. *From* ... **Long system life**	*To* ... **Faster refresh**

Figure 5. Overarching themes that will guide Air Force S&T during 2010-2030 to accelerate delivery of technologies appropriate to the changing spectrum of threats. Emphasis should decrease – but not end – in research supporting the areas at left to accommodate increased emphasis in those on the right. These trends represent shifts in research emphases, not necessarily in near-term acquisition priorities.

3. From … Fixed to … Agile

The Air Force is likely to face a broad spectrum of conflict and non-conflict operations in this period. This will demand a shift from systems designed for fixed purposes or limited missions to ones that are inherently agile in their ability to be readily and usefully repurposed across a range of scenarios. In many cases, this is likely to entail "less than optimal" performance for a primary purpose in exchange for acceptable utility and performance across these broader uses.

This extends also to the range of operating environments in which such systems will be used. Frequency agility, for example, is virtually certain to become an essential attribute as further portions of federal or shared spectrum worldwide are transferred to primary commercial use. Similarly, frequency agility may enable approaches such as pulse-to-pulse waveform diversity that can assist in offsetting the threats posed by DRFMs and other electronic countermeasures. A further example of agility is represented by systems in which sensitive functional components can be more easily removed to allow them to be readily shared with international partners, or to assist in overcoming multilevel security issues.

Technologies and enabling design principles will be needed to permit such increasingly agile systems to retain the greatest functionality in their primary role(s) while allowing the greatest range of intended secondary roles and, to the greatest extent possible, adequate functionality even under unanticipatable scenarios. To a degree far exceeding what is possible today, Air Force systems will need to become increasingly multi-role capable. The benefits of agility reach well beyond the scenarios identified above, and will allow many systems to swing from high-end, general-purpose applications to lower-end irregular warfare applications.

For reasons both of cost savings and flexibility of use it will become increasingly necessary to develop technologies specifically intended to enable substantially greater agility in systems.

4. From … Control to … Autonomy

Airborne remote-piloted systems and many non-airborne systems have significant autonomous capability, however increasing levels of "flexible" autonomy will be needed for a wider range of Air Force functions during 2010-2030 and beyond. These will include fully unattended systems with reliable levels of autonomous functionality far greater than is possible today, as well as systems that can reliably make wide-ranging autonomous decisions at cyber speeds to allow reactions in time-critical roles far exceeding what humans can possibly achieve.

Key attributes of such autonomy include the ability for complex decision making, including autonomous mission planning, and the ability to self-adapt as the environment in which the system is operating changes. Flexible autonomy refers to systems in which a user can specify the degree of autonomous control that the system is allowed to take on, and in which this degree of autonomy can be varied from essentially none to near or complete autonomy.

Limited autonomy is already being widely used in Air Force systems, but their autonomous functionality is far below what could be achieved over the next decade. From a purely technical perspective, methods already exist to extend autonomy to far higher levels than are being used today. Even this level of autonomy will be enabled only when automated methods for verification and validation (V&V) of complex, adaptive, highly nonlinear systems are developed. In the near- to mid-term, developing methods for establishing "certifiable trust in autonomous systems" is the single greatest technical barrier that must be overcome to obtain the capability advantages that are achievable by increasing use of autonomous systems.

Although humans will retain control over strike decisions for the foreseeable future, far greater levels of autonomy will become possible by advanced technologies. These, in turn, can be confidently exploited as appropriate V&V methods are developed along with technical standards to allow their use in certifying such highly autonomous systems. This area is at the interface between complex nonlinear systems theory and the massive computational analytics enabled by the availability of near-infinite processing and storage capacity at near-zero unit cost.

It is thus reasonable to expect that with increasing S&T emphasis in this area, technologies will be developed over the next decade that can enable reliable V&V and certification methods to provide "trust" in even highly adaptable autonomous systems. These in turn can open up entirely new avenues for reducing manpower needs, and entirely new missions and capabilities for Air Force systems. Such trusted highly autonomous systems will be essential for dominance in the cyber domain, where operation at the speeds that autonomy offers will be combined with levels of autonomous control far beyond what is possible today.

Note that potential adversaries may be willing to field highly autonomous systems without any demand for prior certifiable V&V. In so doing they may gain potential capability advantages that we deny ourselves by requiring a high level of V&V. In general, the enormous capabilities that such highly autonomous systems can provide are one of the main conclusions of this study.

Emphasis on developing methods to achieve verification and validation (V&V) of complex, highly adaptive, autonomous systems will be essential to enabling the capabilities they can provide.

5. From ... Integrated to ... Fractionated

To date, in most system architectures the various functions needed for the system to operate have been physically integrated via subsystems in the overall system. Thus, for example, most aircraft integrate communications, electronic warfare, ISR, strike, and other mission functions together into a single platform. The resulting systems become relatively large and heavy, with associated performance penalties that translate into limits on range and other factors, and with relatively high unit production costs and operating costs. Moreover, in such system architectures the loss of any one subsystem can lead to mission failure or even loss of the entire system. From a mission survivability perspective, such architectures demonstrate poor survivability.

Yet it is becoming increasingly possible to "fractionate" systems into a relatively small set of functional subsystems, each operating spatially separated from the others but maintaining local communication with them to allow overall system functionality to be preserved. The system is, in effect, separated into fractional elements and physically dispersed. If the fractionation is done correctly then the short-range communication bandwidth between elements can be made low enough that burst mode transmission, frequency agility, laser links, and other methods can be used to achieve low probability of detection and maintain jam-resistant local link integrity.

Fractionation provides a dispersed architecture that, even by itself, can create cost and risk advantages over traditional integrated systems. It is distinct from modularity in that the latter is designed to more readily allow subsystem replacement but still aggregates subsystems into an integrated architecture. Most importantly, when a fractionated architecture is augmented by even low levels of redundancy among the dispersed elements, survivability can increase dramatically. Combining fractionation with redundancy in a system-level architecture produces survivability that far exceeds what is possible by traditional redundancy alone.

For a system architecture with *m* different types of functional elements and *n* copies of each, the number *N* of shots each having individual probability-of-kill *Pk* that are needed to cause mission-loss probability *PL* can be readily shown to be given by

$$N = \frac{\log\left(1-\left[1-\left(1-P_L\right)^{1/m}\right]^{1/n}\right)^{mn}}{\log\left(1-P_k\right)}$$

Any fractionated architecture corresponds to particular *m* and *n* values, and the traditional integrated system architecture corresponds to the trivial case where $m = n = 1$. Using the above equation it can be readily shown that, for reasonable P_k values, the number of shots N needed to cause any mission-loss probability P_L increases rapidly with the degree of fractionation and redundancy. The combination of fractionation and redundancy produces a multiplicative benefit beyond what is possible in a traditional architecture.

Such highly survivable fractionated system architectures are only recently becoming possible through technologies that enable jam-resistant secure local communications and collaborative control to permit the spatially dispersed functional elements to operate as a single coherent system. Fractionated architectures are a key to the development of low-cost autonomous systems that can survive in anti-access/area-denial environments. Such architectures and the key technologies that enable them have been relatively unexplored to date.

Fractionated systems represent a relatively new technology-enabled architecture for achieving increased system-level survivability with low-cost autonomous system elements.

6. From … Preplanned to … Composable

Composability refers to the ability to rapidly assemble an as-needed mission-specific capability from an available set of standard and relatively low-cost elements. It raises the concept of agility from the level of a single system to the entire set of systems needed for mission-level capability. By composing a mission-level capability from a standard set of common elements, a broader range of missions can be effectively addressed at lower cost than is possible with a more traditional aggregated system designed for a specialized or narrow range of missions.

For example, the same fractionated mission elements noted above could be used to compose a relatively simple EO surveillance mission in uncontested airspace using just one element having imaging capability. The same set of elements could be used to compose a hunter-killer mission package by adding the strike variant. At the other extreme, for heavily defended airspace a large number of EO, EW, strike, and other variants could be composed into a package suitable for that mission. The key point is that for a wide range of mission types, it may be possible to compose a "good enough" mission package from the appropriate mix of a set of common-core flexibly-autonomous elements that differ primarily in their function-specific payloads.

The underlying concept of composability can be applied broadly. In cyberspace, for example, by storing elements from a common set of functional cyber-mission components in a network in advance of a cyber attack, these elements can be very rapidly composed into a "good enough" mission package to respond to the attack. The point is that composing a mission package from an already-available set of common elements can allow for a faster response at lower cost that is "good enough" to deal with a much wider range of missions.

In the cyber domain, the fundamental inability to anticipate every possible attack and preplan an ideally tailored response to each is readily apparent. However the same idea of a mission package composed as needed from pre-existing low-cost elements is central to the previous example as well. Similarly, plug-and-play modularity to enable rapid custom satellite composition is a further embodiment of composability, and the approach can even be extended to as-needed composition of "good enough" small satellite constellations for operationally responsive space missions.

Gaining the cost advantages and mission flexibility that are available through composability will require developing technologies to enable composable mission capabilities in all domains.

7. From ... Single-Domain to ... Cross-Domain

Air Force systems are broadly thought of as belonging in the air, space, and cyber domains, yet it is being increasingly recognized that key vulnerabilities can arise in these systems as a result of cross-domain influences and interdependences. It is possible to exploit cross-domain effects in adversary systems as well, or even to derive beneficial effects from cross-domain influences. Here, cross-domain effects are defined to include a technology in one domain that can produce unexpected beneficial or detrimental effects in another domain, or a technology in one domain that requires supporting functions in another domain and as a result creates interdependences between two or more domains, or even a technology that falls "between" classical domains but has implications in one or more of them.

In general, the awareness of these cross-domain effects is only beginning to be understood, in part as a result of recent rapid growth in vulnerabilities of cyber and space systems. Cross-domain effects apply, however, in all three domains. Over the next decade, efforts to develop new technologies will need to increasingly take possible unintended cross-domain effects into account. At the same time, technologies to minimize existing cross-domain vulnerabilities, and to exploit similar vulnerabilities in adversary systems, will be increasingly needed.

Emphasis will increase on technologies to understand, anticipate, and where appropriate avoid cross-domain effects in Air Force systems, and to exploit them in adversary systems.

8. From ... Permissive to ... Contested

With only a few exceptions, Air Force operations over the past two decades have been conducted under largely permissive conditions in relatively poorly defended airspace. This risks the development of an under appreciation for the impact that possible future operations in anti-access/area-denial environments could have on some missions and functions, including those in the cyber domain. Greater emphasis should be placed on technologies that can operate effectively in moderate or heavily contested domains.

This will require developing technologies specifically designed to enable resilience of key Air Force mission functions in contested domains. In developing technology solutions for any specific problem, it will be essential to ensure that these have the needed resilience to operate effectively even in significantly contested domains. Examples include the need for electronic warfare technologies that can be effective in situations where control of the electromagnetic domain is not assured. Similarly, technologies will be needed that can permit remotely-piloted aircraft to communicate and operate under threats in defended airspace, and that can permit other weapon systems to achieve greater range and deep penetration ability. In the increasingly contested space domain, technologies will be needed to ensure resilience of space assets and space operations to various forms of threats.

To meet the needs of Air Force operations during this period, it will be essential for S&T emphasis to increase on technologies that can remain effective even in contested domains.

9. From ... Sensor to ... Information

As a result of rapid advances made in electronics and in imaging technologies in particular, there has been an enormous surge in the amount of sensor data available across many portions of the electromagnetic spectrum. Current EO/IR sensors, for example, generate huge amounts of raw imaging data, and similar increases in raw data streams are occurring from other sensing modalities, including hyperspectral imaging, LIDAR imaging, electronic signals intelligence, and other RF-based ISR sensing modes. The volume of sensor data from current-generation sensors that must be processed, exploited, and disseminated has become overwhelming, as manpower requirements to deal with these data have placed an enormous burden on the Air Force.

It appears clear that the focus on developing ever more capable sensor technologies of the present type is far outstripping the ability to deal efficiently with the resulting raw data. Such current-generation sensors can be characterized as "dumb" insofar as they are designed solely to generate raw sensed data, rather than to generate interpreted information from these raw data. Over the next decade, development focused solely on higher-output sensor technologies will need to be de-emphasized to accommodate increased emphasis on developing technologies for "intelligent" sensors. These will, for example, provide cueing-level processing of raw data on the sensor itself and then transmit only those data segments containing cues to the ground.

Technologies that make use of continuing advances in miniaturization of electronics and the availability of massive processing and storage in extremely small volumes can in principle allow limited processing of raw sensor data to be done behind the sensor itself. This offers potential to transform sensor output from raw data to real information. Processing done on the sensor need not be as detailed as that currently done on the ground, and might be used only to cue burst transmissions of raw data to the ground for subsequent detailed processing. Resulting substantial decreases in the amount of raw data being transmitted would alleviate bandwidth requirements as well as ground-based PED manpower requirements.

Beyond development of such "intelligent sensors" that provide information rather than data, other technologies may assist with increasingly autonomous data processing on the ground to reduce current PED manpower requirements. As with intelligent on-sensor processing, such autonomous off-sensor processing may not need to be as detailed as current manual processing, since it might only be used to cue raw data segments for subsequent manual processing. The combined impacts of "intelligent sensors" to reduce raw data streams and autonomous ground processing to assist in current PED functions can potentially be enormous.

In general, Air Force S&T emphasis in the next decade will need to shift focus from advancing current "dumb" sensor technologies to advancing the ability to extract useful information from the sensor data stream. The utility of ever-greater situational awareness created by expanded ISR systems will make this an exceptionally high priority for the S&T community.

A significant shift in Air Force S&T emphasis is needed toward developing technologies that can increasingly assist in converting raw sensor data into useful information.

10. From … Strike to … Dissuasion/Deterrence

Dissuasion and deterrence have played a role in the Air Force mission since its inception as a distinct Service. However, the nature of the broader deterrence mission that applies in today's environment, and that is likely to become increasingly important during 2010-2030, is substantially different. These include nuclear and conventional components, and must address state and non-state actors as well as other entities that are not dissuaded or deterred y traditional means based solely on threats of retaliation or punishment. The corresponding tools needed to effectively deal with that environment are also different. In addition to technologies needed for active conflict operations, there will be a need during this period to enable appropriate dissuasion and deterrence tools for that environment.

As noted in Section 2, deterrence in the context of 2010-2030 involves not only effectively deterring nations that are pursuing nuclear-weapons capabilities, such as North Korea and Iran, but also deterrence in the conventional warfare sense, as well as in the cyber domain and even deterrence of chemical and biological warfare arenas. The broad spectrum of the Air Force role in these missions, and the wide range of actors involved, creates a need for significantly different and more effective tools that can support these deterrence roles.

In many cases, to be effective these tools will require non-kinetic means to exert influence on potential adversaries. The cyber domain offers significant capabilities for exerting a deterrent influence on some such actors. Deterrence in the context of what is often termed "irregular warfare" and "hybrid warfare" brings special difficulties that may require new approaches. Actors other than nation states may be less susceptible to such influence, and other tools will be needed to deal with them.

In general, S&T will need to play an expanded role in developing technologies that can support the Air Force role in providing effective dissuasion and deterrence capabilities across the spectrum of conflict and non-conflict scenarios. These efforts may benefit from coordination with the other Services, with DARPA and IARPA, and with our international partners.

Over the next decade, technologies will be needed that can enable effective deterrence capabilities across a far broader spectrum of threats than has historically been the case.

11. From … Cyber Defense to … Cyber Resilience

Since the earliest awareness of the vulnerabilities of cyber systems and the far-reaching implications of them, efforts to control these have been based largely on "cyber defense". This refers to the many traditional approaches intended to prevent adversaries from entering computer networks and other cyber-accessible systems. Yet despite extensive efforts to block adversaries from entering these systems, they can and routinely do access them. Today, the efforts that must be expended to protect cyber systems in this manner far exceed those that adversaries expend to penetrate them, and there is essentially no evidence to suggest that this will change significantly as long as the focus remains largely on cyber defense.

There will be a need for entirely new approaches to assuring continued mission effectiveness of cyber systems, allowing us to better "fight through" cyber attacks. A key step is to begin a fundamental philosophical shift from emphasizing "cyber protection" to emphasizing "mission assurance". This subtle but important change opens entirely new ways to approach the cyber challenge. For example, mission assurance can be dramatically increased by technologies that can inherently make cyber systems more resilient to the presence of adversaries within them. S&T efforts to develop technologies for increased "cyber resilience" will be a key theme in the coming decade, and one of the cornerstones for achieving mission assurance.

For instance, as described in Section 5.2.1, massive virtualization coupled with rapid network recomposability can make cyber systems inherently far less vulnerable to direct network attacks. This technology represents a fundamental shift in the approach to network defense, replacing current efforts to keep adversaries out with a new paradigm that makes it nearly useless for an adversary to gain entry. The resulting network thereby becomes inherently resilient to attack.

Such inherent "cyber resilience" via massive virtualization and rapid network recomposability can also cause cyber adversaries to leave behind far greater forensic evidence for attribution, regardless of whether they entered via the network layer, backdoor implants in hardware, or the application layer. The technology can provide inherent "resilience" in the sense that its benefits do not require any explicit awareness of the adversary's presence in the system or explicit action by the defender to gain the benefit.

The fundamental shift in focus from cyber defense to cyber resilience represents one of the most far-reaching themes for Air Force S&T in the coming decade.

12. From … Long System Life to … Faster Refresh

It is a reality that many of the air platforms and other systems that the Air Force employs will continue to be needed and remain in use long after they are first introduced into operation. Yet such long-life systems often make integrating new technology-derived updates into them inherently difficult and costly. This is particularly true when systems are not developed with architectures specifically designed to accommodate such future upgrades. At the same time, technologies continue to advance at an increasingly faster pace, and worldwide access to the resulting military systems developed from them is also increasing rapidly.

The result is that the global "technology refresh rate" is now far faster than just a decade or two ago. Air Force capabilities based on systems designed for long-life use will remain appropriate for some applications, however adversary capabilities obtained with less exquisite systems may have increasing access to faster technology refresh cycles. For many types of systems, it will become increasingly difficult for the Air Force to maintain sufficient technology advantage without itself having access to substantially faster technology refresh rates.

Technologies need to be developed to allow subsystems, if not entire systems, to be inherently "expendable by design" in the sense that they are designed and integrated with the specific intent to replace them with newer technology after far shorter usage than is the case today. In part, this requires new technologies that can enable key system functions to be implemented at much lower cost. It also requires advances in technologies for "open architectures" that are secure but can enable ready integration of successive generations of newer subsystems.

Development of technologies to support such innately expendable systems also supports other overarching themes identified in "Technology Horizons", including agility and composability, as well as the shift toward fractionated system architectures. All of these are essential elements for maximizing U.S. Joint capabilities while supporting the need for Air Force and DoD to move from exquisite systems to more flexible, agile, and broadly effective capabilities that can meet the needs of a wider and increasingly less predictable set of future operational scenarios.

New approaches that can enable far faster technology "refresh rates" in Air Force systems and subsystems are key to achieving the far greater flexibility that will be needed to respond to the range of future threats.

SECTION 5. TECHNOLOGY-ENABLED CAPABILITIES FOR THE AIR FORCE IN 2010-2030

Potential technology-enabled capability areas are identified that are aligned with key needs of the strategic environment, enduring realities, and overarching themes that the Air Force faces, and that can be achieved with realistically attainable technology advancements over the next ten years.

The notional process used in "Technology Horizons" for identifying "potential capability areas" is shown in Figure 6. For each of the air, space and cyber domains, it identified key functions associated with the domain, and used these to envision potential capability areas (PCAs) that could provide these domain functions. For each candidate PCA, it considered the technology areas that would be needed to enable that capability and the state to which each technology area would need to be brought in order to provide the potential capability being envisioned.

If a technology area could not be credibly advanced from its current 2010 state to the needed state by the 2020 target in the "10+10 Technology-to-Capability" process in Figure 2, then an alternate technology for enabling that PCA had to be identified or the capability was rejected. Similarly, if an adversary using the same state of technologies could readily counter a potential capability, or if a capability would create unacceptable vulnerabilities -- either by itself or through cross-domain influences it would produce -- then it was also rejected. In so doing, "Technology Horizons" envisioned potential capability areas that would be of value in the 2030 strategic environment and could be credibly achieved if the needed S&T investments are made.

Note this does not mean that the technologies needed to develop these capabilities are already sufficiently mature, or the Air Force should necessarily acquire these capabilities. Instead, these notional capabilities are identified because they are well matched to needs of the Air Force within the time horizon of the study. This set of potential capability areas is then used as a mechanism for identifying the key technology areas (KTAs) that would be needed to enable them. In Section 6, mapping the KTAs across the PCAs allows identification of the most valuable technology areas that Air Force S&T should be pursuing now.

This process allowed "Technology Horizons" to maintain its focus on assessing what the most valuable technology areas for the Air Force are, without specifying what future capabilities the Air Force should be acquiring. Note that the "potential capability areas" used in this process should not be interpreted as an alternative to "flagship capability concepts" or "focused long-term challenges" that AFRL may use as constructs for communicating, planning, and managing its S&T efforts. Instead, the potential capability areas are a complementary means used here for understanding the relative impacts of various key technology areas. They allow the KTAs to be identified that have the greatest crosscutting value in enabling capability areas that are aligned with the strategic context and enduring realities of 2010-2030.

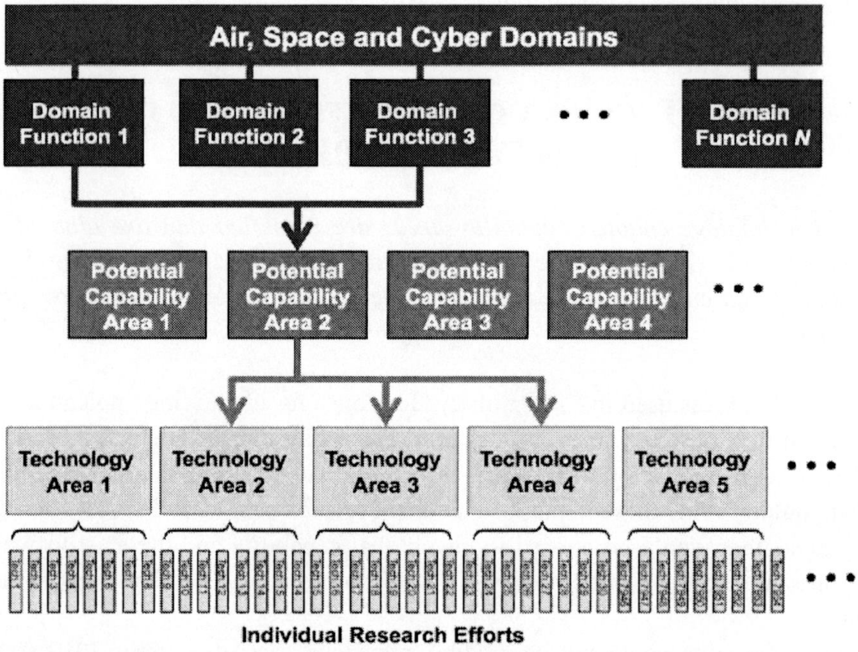

Figure 6. Notional process used in "Technology Horizons" for identifying "potential capability areas" (PCAs) and "key technology areas" (KTAs) for the air, space and cyber domains.

1. Alignment of Capability Areas with Air Force Core Functions

The process described above is predicated on the set of PCAs being sufficiently broad to address most of the key functions the Air Force needs to perform its mission. This ensures that the KTAs that were determined by this process as being the most valuable for Air Force S&T to pursue can support a sufficient range of the functions needed to meet likely Air Force needs. To assess if this is the case, the final set of PCAs was mapped into the Air Force service core functions to determine the range of these that the underlying key technology areas (KTAs) will support.

1.1. Air Force Service Core Functions (SCFs)

The Air Force has defined the following set of 12 "service core functions" that encompass the combat and support functions it needs to accomplish its range of missions.

- SCF1: Nuclear Deterrence Operations
- SCF2: Air Superiority
- SCF3: Space Superiority
- SCF4: Cyberspace Superiority
- SCF5: Global Precision Strike
- SCF6: Rapid Global Mobility
- SCF7: Special Operations
- SCF8: Global Integrated ISR
- SCF9: Command and Control
- SCF10: Personnel Recovery
- SCF11: Building Partnerships
- SCF12: Agile Combat Support

In the next subsection, the final set of 30 "potential capability areas" (PCAs) that were used in "Technology Horizons" are listed, and in the following subsection these are mapped across the above service core functions to assess the range of these that they support.

Mapping the "potential capability areas" (PCAs) across the "service core functions" (SCFs) ensures that "Technology Horizons" considered a sufficiently wide range of technologies to support the full set of Air Force missions.

1.2. Technology-Enabled Potential Capability Areas (PCAs)

A set of 30 "potential capability areas" (PCAs) was identified as being credibly achievable within the time horizon addressed by this study and as meeting key needs defined by the strategic environment, enduring realities, and overarching themes that apply to 2010-2030. Each of these PCAs is listed below, and all are summarized in a following section.

- PCA1: Inherently Intrusion-Resilient Cyber Systems
- PCA2: Automated Cyber Vulnerability Assessments and Reactions
- PCA3: Decision-Quality Prediction of Behavior
- PCA4: Augmentation of Human Performance

- PCA5: Advanced Constructive Discovery & Training Environments
- PCA6: Trusted, Adaptive, Flexibly-Autonomous Systems
- PCA7: Frequency-Agile Spectrum Utilization
- PCA8: Dominant Spectrum Warfare Operations
- PCA9: Precision Navigation/Timing in GPS-Denied Environments
- PCA10: Next-Generation High-Bandwidth Secure Communications
- PCA11: Persistent Near-Space Communications Relays
- PCA12: Processing-Enabled Intelligent ISR Sensors
- PCA13: High-Altitude Long-Endurance ISR Airships
- PCA14: Prompt Theater-Range ISR/Strike Systems
- PCA15: Fractionated, Survivable, Remotely-Piloted Systems
- PCA16: Direct Forward Air Delivery and Resupply
- PCA17: Energy-Efficient Partially Buoyant Cargo Airlifters
- PCA18: Fuel-Efficient Hybrid Wing-Body Aircraft
- PCA19: Next-Generation High-Efficiency Turbine Engines
- PCA20: Embedded Diagnostic/Prognostic Subsystems
- PCA21: Penetrating Persistent Long-Range Strike
- PCA22: High-Speed Penetrating Cruise Missile
- PCA23: Hyperprecision Low-Collateral Damage Munitions
- PCA24: Directed Energy for Tactical Strike/Defense
- PCA25: Enhanced Underground Strike with Conventional Munitions
- PCA26: Reusable Airbreathing Access-to-Space Launch
- PCA27: Rapidly Composable Small Satellites
- PCA28: Fractionated/Distributed Space Systems
- PCA29: Persistent Space Situational Awareness
- PCA30: Improved Orbital Conjunction Prediction

A much larger set of potential capability areas was considered, however this list gives the 30 PCAs that were assessed as being most valuable for meeting Air Force needs. The list is not prioritized in any way, since it was not the objective of the study to define capabilities the Air Force should acquire, but instead to prioritize the most essential technology areas that are needed to enable capabilities such as these. These are potential "capability areas" in the sense that most represent systems or functions that enable what would be regarded as future "capabilities". These capability areas also serve as example embodiments of the broader overarching themes identified in Section 4.

1.3. Alignment of Potential Capability Areas with Service Core Functions

Figure 7 shows the mapping of each of the potential capability areas (PCAs) listed above into each of the 12 Air Force service core functions (SCFs) listed on the previous page. In each case, a check mark indicates that the potential capability area could significantly assist in the performance of at least one or more aspects of the corresponding service core function. Where there is no check mark there may still be connections between that capability area, but these do not rise to a comparably meaningful as those that are marked in the figure.

	SCF #1:	SCF #2:	SCF #3:	SCF #4:	SCF #5:	SCF #6:	SCF #7:	SCF #8:	SCF #9:	SCF #10:	SCF #11:	SCF #12:
PCA #1:	✓	✓	✓	✓	✓	✓	✓	✓	✓	✓	✓	✓
PCA #2:	✓	✓	✓	✓	✓	✓	✓	✓	✓	✓	✓	✓
PCA #3:	✓	✓	✓	✓	✓	✓	✓	✓	✓	✓	✓	✓
PCA #4:	✓	✓	✓	✓	✓	✓	✓	✓	✓	✓		✓
PCA #5:	✓	✓	✓	✓	✓	✓	✓	✓	✓	✓	✓	✓
PCA #6:		✓	✓	✓	✓	✓	✓	✓	✓	✓		✓
PCA #7:	✓	✓	✓	✓	✓	✓	✓	✓	✓	✓	✓	
PCA #8:	✓	✓	✓	✓	✓		✓	✓	✓	✓		
PCA #9:	✓	✓		✓	✓	✓	✓	✓	✓	✓		
PCA #10:	✓	✓	✓	✓	✓	✓	✓	✓	✓	✓		
PCA #11:	✓	✓		✓	✓	✓	✓	✓	✓	✓	✓	
PCA #12:	✓	✓	✓	✓	✓		✓	✓		✓		
PCA #13:		✓		✓			✓	✓				
PCA #14:					✓			✓				
PCA #15:		✓			✓	✓	✓	✓				
PCA #16:						✓	✓			✓	✓	
PCA #17:						✓					✓	
PCA #18:		✓				✓		✓				
PCA #19:	✓	✓			✓	✓		✓		✓		
PCA #20:	✓	✓	✓			✓						
PCA #21:	✓	✓		✓	✓							
PCA #22:	✓	✓			✓							
PCA #23:		✓			✓		✓				✓	
PCA #24:		✓			✓		✓				✓	
PCA #25:					✓		✓					
PCA #26:			✓					✓				
PCA #27:			✓		✓		✓	✓				
PCA #28:			✓					✓				
PCA #29:			✓		✓			✓				
PCA #30:			✓								✓	

Figure 7. Mapping of "potential capability areas" [PCAs] across Air Force service core functions [SCFs].

2. Brief Descriptions of Technology-Enabled Capabilities

Each "potential capability area" is briefly described below at a level sufficient to understand the basic capabilities being proposed, the technologies that can enable them, and the benefits they would provide for the Air Force. For each of these PCAs, there is technical support for the claim that the underlying technologies that can be brought to the needed level of readiness by the 2020 technology horizon date in the "10+10 Technology-to-Capability" process.

2.1. Inherently Intrusion-Resilient Cyber Systems

This represents a fundamental shift in emphasis from "cyber protection" to "maintaining mission effectiveness" in the presence of cyber threats, using technologies such as IP hopping, network polymorphism, massive virtualization and rapid network recomposition that can make cyber systems inherently resilient to intrusions entering through the network layer. These convert the currently static network layer into a highly dynamic one, in which the hypervisor mapping between the hardware and functional layers changes constantly in a pseudo-random way, perhaps hundred of times every second. A cyber adversary who finds vulnerabilities in the physical layer thus has virtually no time to use them for mapping the network before its topology has changed. Current manpower-intensive cyber protection efforts to block attacks at the network surface could be augmented or replaced by these new methods, which introduce "rapid maneuver" to cyber warfare to make a network inherently resilient to attack. Since adversaries entering or passing through the network layer may have only a few milliseconds to operate, not weeks or months, they leave greater forensic evidence for attribution. Intrusion-resilient methods would initially be applied to computer networks, but are extendible to other cyber systems.

Commercial development of massive virtualization and hypervisor technologies to enable cloud computing can provide much of the technology needed for such inherently intrusion-resilient Air Force cyber systems.

2.2. Automated Cyber Vulnerability Assessments and Reactions

Assuring continued mission effectiveness in the cyber environment begins with identifying mission essential functions across the Air Force enterprise and prioritizing these in terms of individual and interdependent mission impacts. Mapping the dependence of each function over the cyber domain gives insights into critical hidden interdependences that enable autonomic generation of appropriate reactions to attacks at the "speed of cyber" and thereby maximize mission effectiveness under any threat. This is beginning to be done today via a slow manual process for limited and specific threats; entirely new automated approaches can allow scale-up across the entire Air Force cyber enterprise and automatic development of optimal courses of action for any momentary threat set across the enterprise to maintain maximum mission effectiveness. Software tools to autonomously probe and map complex interdependences across the cyber enterprise are still in relative infancy, and tools needed to visualize and understand the resulting cyber vulnerabilities are primitive. Autonomous decision support tools will also be needed to allow reactions at speeds needed to limit the effects of cyber attacks. Such automated mapping of cyber vulnerabilities and corresponding attack responses, as suggested in Figure 8, represent one way of implementing the overarching theme of "resilience" in Air Force cyber systems.

Figure 8. Automated tools to map mission dependences and resulting vulnerabilities in the cyber domain allow autonomic reactions at the "speed of cyber" to maximize mission effectiveness.

Cyber defense must shift emphasis from network protection to maintaining enterprise-wide mission effectiveness; automated methods for continuous assessment of interdependences among cyber elements and generation of autonomic reactions to threats will be key.

2.3. Decision-Quality Prediction of Behavior

Advances in human and cultural behavior modeling, social network modeling, cognitive modeling, and autonomous reasoning can enable decision support tools for anticipating and predicting adversary and own-force behaviors. Fusion and understanding of information from disparate sources, including classical intelligence data as well as large data sets of open source information routinely collected from global cyber networks, provides essential inputs. Massive analytics made possible by the availability of near-infinite processing and storage capacities during this time can provide software tools that aid in decision-making, COA development, and related tasks involving prediction of individual/collective behaviors over a range of conditions, along with statistical uncertainty bounds to quantify decision confidence levels and inform alternative courses of action.

Massive storage and processing power during this time will enable wider use of open-source and other information and advanced statistical models of individual or collective behaviors.

2.4. Augmentation of Human Performance

Human performance augmentation will be essential for effectively using the overwhelming amounts of data that will be routinely available during this time. As suggested in Figure 9, this may include implants, drugs or other augmentation approaches to improve memory, alertness, cognition and visual/aural acuity. It may even extend to limited direct brainwave coupling between humans and machines, and screening of individual capacities for key specialty codes via brainwave patterns and genetic correlators. Adversaries may use genetic modification to enhance specific characteristics or abilities. Performance augmentation will find routine use in the cockpit, on the flight line, by ISR operators and by commanders. Data may be fused and delivered to humans in ways that exploit synthetically augmented intuition to achieve needed decision speeds and enhance decision quality. Human senses, reasoning, and physical performance will be augmented using sensors, biotechnology, robotics, and computing power.

As natural human capacities become increasingly mismatched to data volumes, processing capabilities, and decision speeds, augmenting human performance will become essential for gaining the benefits that other technology advances can offer.

2.5. Advanced Constructive Discovery & Training Environments

Constructive environments being used for training today are simple by the standards of commercial massively multi-player online games (MMOGs), many of which have hundreds of thousands of users around the world playing at any given moment in thousands of simultaneous multi-player games; CounterStrike™ alone has over 4.5 billion man-minutes of gameplay worldwide every month. Advances in such highly constructive environments – compared with virtual systems in Figure 10 that are in Air Force use today --enabled by continued processing and storage capacities can provide sufficient fidelity to allow useful mining of the enormous statistical information contained in such global gameplay. This can enable entirely new methods for discovering cultural insights as well as innovative blue-and red-force concepts of employment long before they are evident by ordinary experience or in the far smaller statistical samples available by formal methods. Highly constructive environments also enable further improved training at reduced cost, and will be used for high-fidelity mission rehearsal, COA development, and other uses.

The fidelity of constructive environments will benefit from commercial advances in processing, storage, and advanced authoring tools developed for online game markets; expanding the use of such environments offers substantial new capabilities as well as cost savings.

2.6. Trusted, Adaptive, Flexibly-Autonomous Systems

Inexpensive processing power, data storage and powerful autonomous reasoning algorithms will enable systems capable of far higher levels of autonomy and extension of autonomous systems to entirely new functions. Such advanced autonomous systems will be powerful force multipliers and enable operations at timescales far faster than possible with human-in-the-loop control, as suggested in Figure 11. Establishing trust in autonomy will thus become the central factor in gaining access to the potentially enormous capabilities that such systems can offer. Human control will be preserved with legally reviewed rules of engagement and appropriate human-on-the-loop decision points, however adversaries are unlikely to be constrained by such limits and may be able to field systems that can take advantage of far greater levels of autonomy. Entirely new approaches will be needed to enable verification and validation (V&V) for the near-infinite-state control systems created by high levels of adaptability and autonomy. Developing such methods will be essential for gaining certifiable trust and legal freedom to field autonomous systems that can match or exceed potential adversary capabilities.

Methods to enable trust in highly adaptive autonomous systems are a game-changing technology needed to reap the enormous capability and cost benefits that such systems can offer, and to avoid the asymmetric advantage that adversaries could otherwise have with such systems.

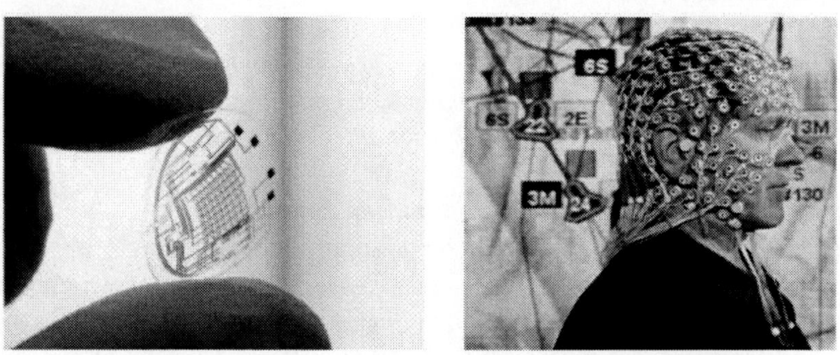

Figure 9. Technologies to augment human performance and cognition have significant potential for reducing Air Force manpower needs or increasing effectiveness in specialized roles.

Figure 10. Virtual and constructive technologies can be advanced to allow high fidelity, constructive, massively multiplayer environments for discovery, including online environments where gameplay statistics are mined to discern cultural insights and identify potential concepts of employment.

Figure 11. Manpower-intensive Air Force functions performed using trusted autonomous systems can enable substantial manpower efficiencies and provide time-domain operational advantages over adversaries limited to human planning and decision processes.

2.7. Frequency-Agile Spectrum Utilization

Future Air Force systems will need substantially frequency-agile capabilities to ensure access to available spectrum bands in worldwide operations. Spectral agility will also be needed for jam-resistance, low-probability-of-detection/intercept, and cyber resilience in the increasingly congested spectrum environment and increasingly contested electronic warfare environment. Yet even software-defined spectral agility has been difficult to implement and certify. New methods such as dynamic spectrum access and associated policy engines will require compatibility with Air Force-specific aspects of spectrum use, including fast movers, large fields of regard, compatibility with below-noise signals such as radars, and resistance to malevolent interferers. Frequency agility can also be extended to spectral mutability, in which a wider range of spectral parameters may adjustable.

> *Frequency agility will be essential for ensuring that Air Force systems can operate under rapidly changing spectrum regulations worldwide, and can provide needed functionality in contested environments where electronic and cyber warfare could create cross-domain vulnerabilities.*

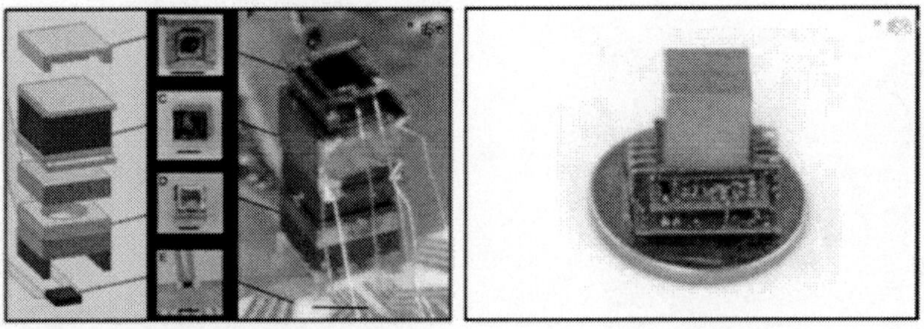

Figure 12. Chip-scale atomic clocks can provide low-drift timing independent of active GPS links, allowing a backup timing source to permit a wide range of critical Air Force systems to continue operating even under GPS-denied conditions.

2.8. Dominant Spectrum Warfare Operations

Today the Air Force relies on U.S. Joint force partners for electronic warfare support, yet electronic warfare will become an essential resident capability for many types of systems. Growing overlaps in electronic and cyber warfare extend this need to full-spectrum dominance, including electronic attack, defense and support, and corresponding functions for optical free-space and waveguided EM systems. In modern IADS environments, jamming may have to overcome sophisticated electronic counter countermeasures in radars and SAM seekers via methods such as cognitive EW. Increasing use of track fusion methods may give adversary systems improved jam resistance, and increased standoff ranges imposed by long-range SAM systems further complicate spectrum warfare. As potential adversaries increasingly move to AESA radars and make use of DRFM-based systems for RF spoofing, Air Force freedom of operation in the air-to-air EM domain is also being significantly challenged. New spectrum warfare methods should be developed to support Air Force dominance in attack, self-defense and support across the EM spectrum.

> *Rapidly advancing EW capabilities of some potential adversary systems may create significant challenges to traditional Air Force spectrum dominance; a new generation of spectrum warfare methods is needed to ensure acceptable freedom of operations.*

2.9. Precision Navigation/Timing in GPS-Denied Environments

Widespread dependence of critical Air Force and other DoD systems on GPS for precision navigation and timing functions has caused potential adversaries to exploit GPS jamming as an asymmetric advantage. Key systems will thus need GPS-independence or augmentation to allow their use in such environments. Chip-scale atomic clocks as shown in Figure 12 and inertial measurement units based on cold-atom principles or other technologies as shown in Figure 13 can provide low-drift PNT in the event of GPS loss. These approaches maintain GPS-like position and timing uncertainties over relatively long periods after GPS signal loss. Netting to systems outside the jammed environment will be able to relay intermittent reference position and timing information for limited updates to correct drift during long-duration jamming events. Technologies can enable miniaturization of such systems and supporting system-level network functions to negate the asymmetric adversary advantage that GPS jamming could otherwise provide.

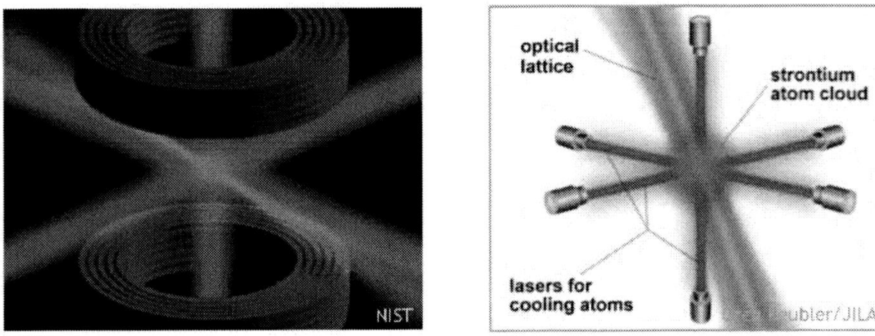

Figure 13. Cold-atom devices enable low-drift IMUs and clocks for PNT in GPS-denied environments. Intersecting lasers trap/cool ytterbium, strontium, or other atoms to slow their motion, forming a Bose-Einstein condensate that can serve as a highly precise matter-wave interferometer.

Developing miniaturized cold-atom IMUs and clocks as well as other approaches for chip-scale PNT will be essential to ensure access to precision navigation and timing in GPS-denied areas.

2.10. Next-Generation High-Bandwidth Secure Communications

Laser communication technology allows extremely high-bandwidth communications over moderate ranges in clear weather conditions for atmospheric propagation, and over very long-ranges when used for crosslinks in space. The extremely low sidelobes produced in transmissions at optical wavelengths also allow for potentially secure communications. When coupled with quantum key distribution (QKD), laser communication can provide verifiably secure encrypted transmissions. These characteristics make laser comms attractive for linking critical Air Force satellite, air, and ground network nodes, and can help shift the rapidly growing demand for bandwidth to frequencies where spectrum management is far simpler. Developing precision pointing and other technologies to enable and integrate laser communications in key Air Force systems will be a high priority.

Technologies to enable laser-based communication links with quantum key distribution methods can provide large increases in bandwidth and security needed in some types of Air Force uses.

2.11. Persistent Near-Space Communications Relays

Flight vehicles and lighter-than-air systems as in Figure 14 can serve as pseudo-satellites capable of operating at near-space altitudes as communication relays in the event of satellite loss. Netting several such relays can provide theater-level coverage and full connectivity to CONUS, with system-level CONOPS providing needed platform protection. These systems hold significant platform challenges as well as electronics, power and thermal management challenges, but may be key to assured communications under the increasingly contested space environment. Systems designed to remain airborne for extremely long durations can allow launch and ascent to be timed when winds are sufficiently weak, enabling structural masses substantially below those of traditional airships designed for lower-altitude use. Significant technical challenges must be overcome in high-altitude materials, lightweight solar cells and energy storage, and thermal management to enable such systems.

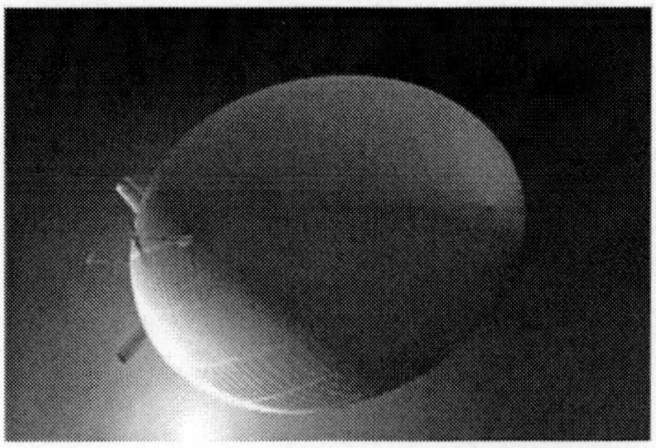

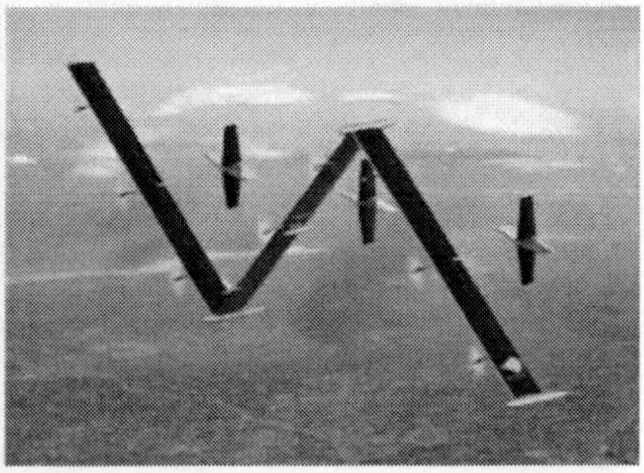

Figure 14. Persistent near-space systems, potentially in the form of ultra-long endurance airships or autonomous flight vehicles, can ensure theater-level communications relay functions in the event of loss or degradation of corresponding space systems.

Near-space communications relays can provide essential backup for satellite communications in the increasingly contested space domain, but represent substantial technical challenges.

2.12. Processing-Enabled Intelligent ISR Sensors

ISR sensors equipped with backplane processing for data synthesis and fusion can permit certain PED functions to be performed on the sensor itself, reducing bandwidth otherwise consumed in transferring large amounts of raw data to the ground. Functions such as coherent change detection may be suited to on-sensor processing, with raw data being transmitted only when cued by processed results from the sensor. On-sensor processing is enabled by the massive processing and storage capacity that can be routinely integrated in modern electronic systems. Processing done on the sensor does not need to be as detailed as can be done on the ground, since its role is only to provide initial cueing that triggers raw data transmission for the ground processing. Beyond substantial reductions in bandwidth consumed by transmission of raw sensor data, intelligent sensors can greatly reduce the manpower needs of current ground-based PED of sensor data.

On-sensor processing of raw data to cue transmission and detailed off-board processing can greatly reduce bandwidth consumed by ISR platforms and also reduce PED manpower needs.

2.13. High-Altitude Long-Endurance ISR Airships

Theater-level ISR aircraft operating for long durations in fixed orbits require forward flight to generate aerodynamic lift, imposing fuel and maintenance costs in addition to engine and aircraft structural masses that cause sensor payload fractions to be far below those of airships. HALE airships designed for extremely long endurance can achieve payload mass fractions even higher than traditional airships, since their ascent through the atmosphere can be timed when winds aloft are low. The size of HALE airships, as shown in Figure 15, accommodates massive sensor apertures far larger than any aircraft could carry, enabling extreme RF sensitivities that may be essential in the emerging strategic environment. Their high operating altitude provides a large field of regard, allowing such systems to also serve as theater-level near-space communication relays. Technical challenges that must be overcome include advancement of high-altitude long-life materials, lightweight solar cells and energy storage, and lightweight multi-functional sensor structures.

HALE airships can provide theater-level ISR capabilities with massive sensors, and offer inherently low operating costs compared with traditional aircraft-based systems.

2.14. Prompt Theater-Range ISR/Strike Systems

Airbreathing hypersonic systems designed for Mach 6 or below avoid most of the extreme temperatures traditionally associated with hypersonic flight, yet can provide rapid ISR or strike capability to engage high-value, time-critical targets from standoff distances. Such systems can hold targets at risk that might otherwise be immune, potentially altering the calculus of strategic deterrence. The system in Figure 16, for example, uses combined-cycle propulsion with internal rocket boost, ram/scramjet acceleration, and scramjet cruise. Inward turning inlets and a dual flow path design allow high volumetric efficiency, and high cruise speed provides significantly increased survivability. Vertical takeoff using a rocket-based combined cycle gives substantially greater range and payload, while imposing few real limitations on launch sites relative to turbine-based horizontal takeoff systems.

High-speed theater-range ISR/strike systems can provide time-critical mission capabilities in a reusable system suited to ranges and operating conditions of key strategic missions.

2.15. Fractionated, Survivable, Remotely-Piloted Systems

A standardized, low-cost, small- to medium-sized, remote-piloted airframe able to carry mission-specific payloads can enable a fractionated and survivable approach for meeting key needs of missions spanning the spectrum from low-end to high-end operations. An as-needed mission package can be composed from such standard elements carrying different payloads, ranging from single or paired elements for uncontested environments to a dozen or more elements in anti-access/area-denial environments. The underlying airframe that enables such an architecture would use small efficient turbojet propulsion, be capable of fully autonomous takeoff, flight, landing, and collaborative control among elements of the mission package, and would carry standard-sized modular payloads such as ISR, EW, strike, communications and other functions among the individual elements to compose the mission capability. Elements cooperate autonomously to form a coherent system using secure, burst-mode, frequency-agile RF or laser communication. Beyond making use of affordable LO technologies, mission survivability in anti-access/area-denial environments would come from the multiplicative benefits of redundancy among the low-cost expendable elements in the fractionated system architecture, which can overwhelm adversary IADS via fundamentally asymmetric cost-imposing strategy.

Fractionated autonomous systems provide a low-cost means for as-needed composable mission capabilities, from individual or paired systems for simple EO/IR imaging or hunter-killer missions to large mission packages with redundancy for anti-access/area-denial environments.

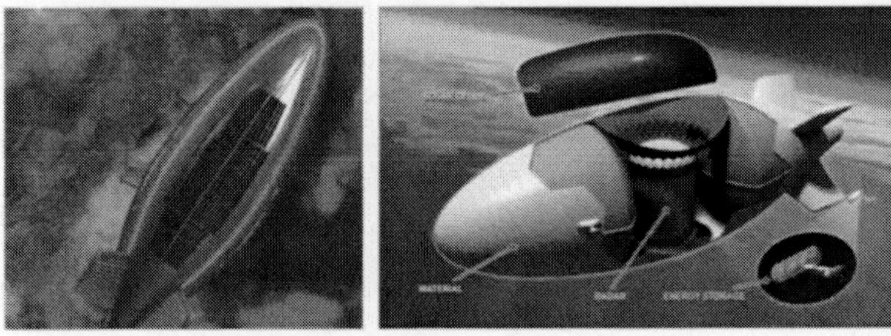

Figure 15. High-altitude long-endurance ISR airships can remain aloft for years, so that launch/ascent are decoupled from immediate demand and thus can be timed when winds are low. This allows exceptionally lightweight structures to be used that can enable extreme functionality.

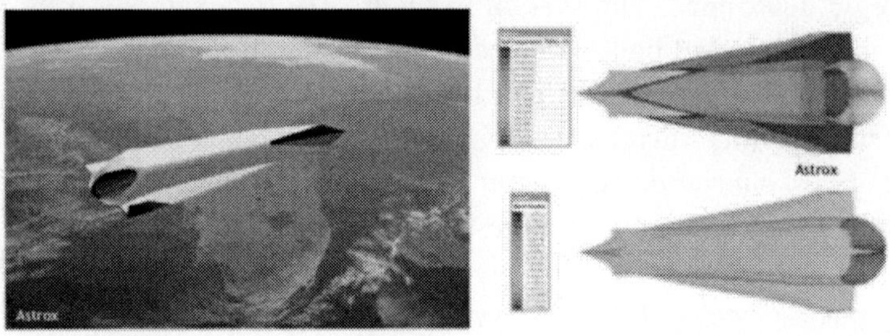

Figure 16. Single-stage ISR/strike vehicles such as this early-stage concept design, which uses internal rocket boost to Mach 3.5 and rocket-based combined-cycle acceleration to scramjet cruise at Mach 6, can enable time-responsive missions at long ranges with substantial payloads and runway landings.

2.16. Direct Forward Air Delivery and Resupply

Small-unit operations in difficult environments are likely to require improved means for resupply beyond current JPADS capabilities and related near-term systems. Such a capability will typically need substantial autonomy and aerodynamic control authority to overcome high winds in mountainous terrain while maintaining precise drop point accuracy, requiring substantially different aerodynamic and control concepts than traditional autonomous systems. Demand for a wide range of weight classes and the ability to deliver from existing mobility platforms make this a substantially challenging system.

The likelihood of a continuing mid- to long-term need to provide precision forward air delivery in difficult terrain makes this an important challenge for advancing autonomous system capabilities within significant limits on cost and other factors.

2.17. Energy-Efficient Partially Buoyant Cargo Airlifters

Partially buoyant cargo airlifters as shown in Figure 17 derive most of their lift from buoyancy and the remainder aerodynamically from forward flight. They typically differ from dirigibles in that they have a rigid semi-monocoque hull structure fabricated from a lightweight but strong composite that provides significant structural mass efficiencies. Moderate-size prototypes of such craft have been built and operated, and fully autonomous systems are possible. Large systems of this type can transport enormous amounts of cargo far more energy efficiently than winged aircraft and more quickly than traditional seagoing cargo ships, and can land in relatively unprepared sites to deliver cargo closer to the point of use.

Enabling such capabilities requires advancing several of the underlying technologies, in system design and integration for differing weight classes and operating capabilities, and in technologies to address weather effects, ballast control, and cargo handling.

Partially buoyant cargo airlifters could provide substantially reduced fuel costs for certain types of air mobility missions and increased ability to deliver large-scale cargo into relatively unprepared sites in a potentially autonomous system.

2.18. Fuel-Efficient Hybrid Wing-Body Aircraft

Subsonic hybrid wing-body aircraft provide significantly increased fuel efficiency and are adaptable for multiple mission roles. Configurations such as in Figure 18 are enabled in part by advanced lightweight stitched composite fabrication methods that allow out-of-autoclave forming of large complex-shaped panels. Hybrid wing-body designs have a wide internal bay suitable for cargo hauling, personnel transport, or as aerial refueling tankers. A single common airframe of this type could be reconfigured during its service life to any of these mission sets to meet changing Air Force mission needs. These designs are suitable to be developed into a family of related aircraft with common wing and other components that can potentially enable reduced production and maintenance costs.

Hybrid wing-body aircraft fabrication is enabled by pultruded rod stitched efficient unitized structure (PRSEUS) technology that can substantially reduce airframe weight and cost.

Figure 17. Hybrid airships that achieve part of their lift from buoyancy and part aerodynamically from forward flight can serve as highly efficient large-scale cargo airlifters for long-haul routes and in uncontested areas with unprepared landing sites.

Figure 18. Hybrid wing-body aircraft have higher fuel-efficiency than traditional tube-and-wing designs for tankers and transports. Advances in integrated stitched-composite manufacturing technologies such as PRSEUS have enabled large, lightweight, high-stiffness panels at low cost.

2.19. Next-Generation High-Efficiency Turbine Engines

Demands for reducing fuel costs and increasing range and endurance call for substantially increased efficiency of embedded turbine engines. Continuing advances in component-level and system-level methods to achieve lower fuel consumption will be needed beyond efforts currently underway, as in Figure 19. These are driving overall pressure ratios well above today's state of the art for increased Brayton cycle efficiencies, in turn requiring new advances in materials and thermal management technologies. At the same time, emerging "third-stream engine architectures" can enable for constant mass flow engines that can provide further reductions in fuel consumption. A wide range of technology advances can enable substantial collective improvements in turbine engine fuel efficiency.

Air Force need for high-efficiency embedded turbine engines in many of its platforms requires substantially different approaches from those accessible to commercial high-bypass engines.

2.20. Embedded Diagnostic/Prognostic Subsystems

An embedded system of diagnostic and prognostic sensors emplaced within high-value Air Force assets such as air vehicles and DE systems will provide warning and prediction of structural or subsystem failure. These systems will ensure preventive maintenance is performed when actually necessary rather than by antiquated time-based maintenance schedules. Data will be autonomously tracked by individual tail/system number and part-by-part. The system can alert maintenance personnel of suspect parts and queue the supply system to ensure replacement prior to failure. These subsystems can also allow maintenance personnel to impose operational flight envelope restrictions on individual tail numbers as needed, rather than imposing these across the entire fleet. Such embedded subsystems can increase overall operational readiness while decreasing maintenance costs.

Embedded sensors and processors to determine system health and maintenance requirements can extend service life of many systems and reduce maintenance and sustainability costs.

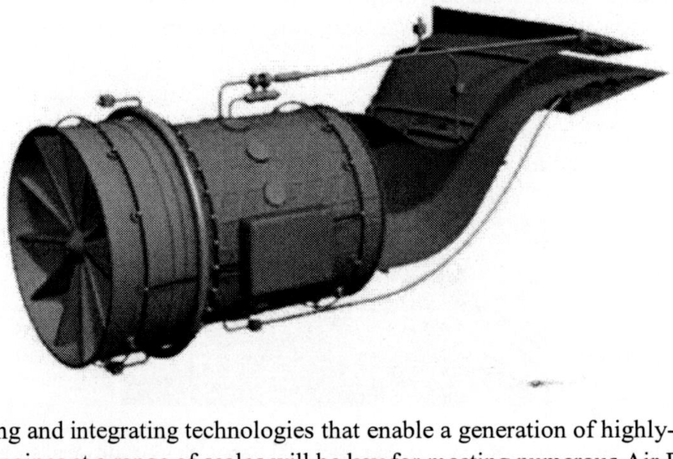

Figure 19. Advancing and integrating technologies that enable a generation of highly-efficient embedded turbine engines at a range of scales will be key for meeting numerous Air Force needs, from fuel cost savings to long-endurance ISR and long-range strike.

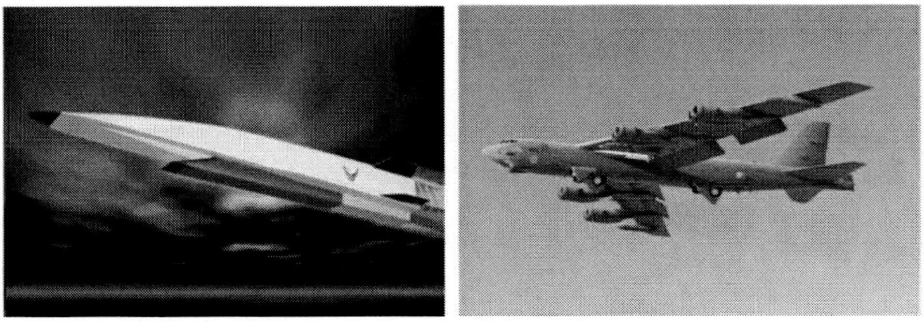

Figure 20. Advances made over the past decade in scramjet propulsion technologies embodied in the X-51 demonstrator can allow development of high-speed cruise missiles and other capabilities to support long range strike on critical targets in anti-access/area-denial environments.

2.21. Penetrating Persistent Long-Range Strike

The need to penetrate substantially defended adversary airspace and persist to achieve desired effects is likely to remain an enduring requirement. This will demand a range of technologies that can achieve deep penetration capability, long loiter, and secure egress. Such systems would likely need LO across multiple bands, secure communications, and advanced active and passive spectrum warfare capabilities to allow deep penetration into adversary territory. Technologies to support efficient engines, bleedless inlets, and serpentine nozzles will be needed for improved propulsion integration to enable long range and loiter.

> *Maintaining low observable capabilities is likely to remain essential for key Air Force missions, as will a wide range of supporting technologies needed to enable long range and persistence.*

2.22. High-Speed Penetrating Cruise Missile

A high-speed cruise missile as suggested in Figure 20 can provide an effective standoff component in a long range strike system. It achieves penetration into anti-access/area-denial environments by using speed to obtain substantially increased survivability against modern air defenses. Such an airbreathing penetrating cruise missile might operate in the high supersonic or low hypersonic regime to provide a balance between flight time, range, and survivability. Advances in dual-mode ram/scramjet propulsion can enable systems up to around Mach 6, while high Mach turbine engine advances would be used for systems operating closer to Mach 3+. Such a system might be air-launched and employ multiple redundant systems for navigation and target acquisition including GPS, chip-scale cold-atom PNT systems, optical terrain recognition, EO and IR sensors, passive RF sensing, and active EW. Variants could include a hard target penetrator or dispenser for sub-munitions.

> *Speed provides an alternative to low-observability that can permit time-critical strikes in anti-access/area-denial environments.*

2.23. Hyperprecision Low-Collateral Damage Munitions

The development of focused lethality munitions, precision guidance via cold-atom PNT systems, and increased ISR information allow development of intelligent, cooperative, small munitions with near-zero CEP capable of identifying and coordinating an attack through simultaneous engagements of a target to achieve a mission-defined level of effect. Multiple munitions could be launched or dropped from one or several air vehicles to cooperatively engage targets. Operating autonomously or with a continuous stream of ISR data, the munitions would do final target acquisition and cooperatively plan a simultaneous attack to achieve the desired effect. Munitions not used would engage other approved targets or self-destruct with minimal collateral damage.

Advances in precision guidance for GPS-denied areas and focused lethality munitions can enable a generation of hyperprecise low collateral damage munitions.

2.24. Directed Energy for Tactical Strike/Defense

Laser and high power microwave systems can be developed with sufficient power and beam control for active airbase defense, air vehicle defense, and tactical strike applications, as suggested in Figure 22. Prior chemical lasers as used in ATL will be replaced by solid-state laser systems, most likely derived from HELLADS technology, and subsequently will be replaced by even more efficient fiber laser systems. The reduced size, weight, power, and thermal management requirements will allow closer integration of such systems in advanced fighters and other tactical platforms. Highly autonomous, self-contained systems can be utilized for self-defense to neutralize credible GRAM and air-to-air missile threats. Airborne tactical strike systems can be used against targets in complex environments such as urban areas with near-zero collateral damage. Such systems can incorporate intelligent beam control to avoid collateral damage to blue forces and non-combatants. Advances in power and thermal management can permit ground-based airbase defensive systems to have near-infinite magazines, while airborne systems will have magazines limited by on board power generation and storage and by thermal management capabilities.

Technologies for laser-based directed energy systems enabling near-zero collateral damage tactical strikes and airborne self-defense are advancing rapidly, and in the mid- to long-term may allow such systems to expand into new airborne and ground-based roles.

Figure 21. India and Russia have developed and fielded the Brahmos supersonic cruise missile reportedly capable of Mach 2.5 flight over ranges in excess of 250 km. Su-30 fighters are reportedly being modified to provide an air-launch capability, primarily for anti-ship and land attack roles.

2.25. Enhanced Underground Strike with Conventional Munitions

Hyperprecision navigation, timing, and autonomous/cooperative guidance will enable two or more conventional munitions to strike high-value underground targets using extremely short and precisely spaced time separations to provide maximal penetration. Using the unsteady dynamics of the compression and expansion wave system produced in granular earth medium by the downward impulse from a preceding munition allows a following time-phased munition to penetrate through the locally-weakened depth-time region produced by the over-expansion wave in the resulting wave system. The second munition in effect "wave-rides" through the subsurface wave system created by the first munition, taking advantage of the degraded material properties of the granular medium that result from partial dislocation of the otherwise interlocking granular constituents. This partial fluidization of the granular medium allows for dramatically reduced drag on the second munition as well as increased penetration depth and stability.

Relatively soft underground targets can be struck without ground penetrating munitions by using precisely time-phased impacts that allow subsurface wave dynamics to be exploited for achieving substantially deeper impact than would otherwise be possible.

2.26. Reusable Airbreathing Access-to-Space Launch

The cost of placing satellites in orbit with current EELV launchers is high. Next-generation two-stage-to-orbit launch systems currently under consideration are focused on reusable rocket approaches that can provide modest reductions in launch costs depending on launch rates. However reusable airbreathing two-stage-to-orbit systems based on combined-cycle propulsion, as in Figure 23, have potential to create large reductions in launch costs, since such systems utilize oxygen in the atmosphere and thus do not need to loft the enormous oxidizer weight needed by rocket systems. Vertical takeoff space launch vehicles such as this will use a reusable rocket first stage with a reusable rocket-based combined-cycle airbreathing scramjet second-stage. Advanced materials, automation, and on-board diagnostic and prognostic systems will allow rapid turn-around for reuse as well as life cycle cost savings as compared to conventional rockets.

Figure 22. Laser directed energy systems will use advanced solid state and fiber laser technologies that enable lightweight, efficient airborne self-defense and low collateral damage tactical strike.

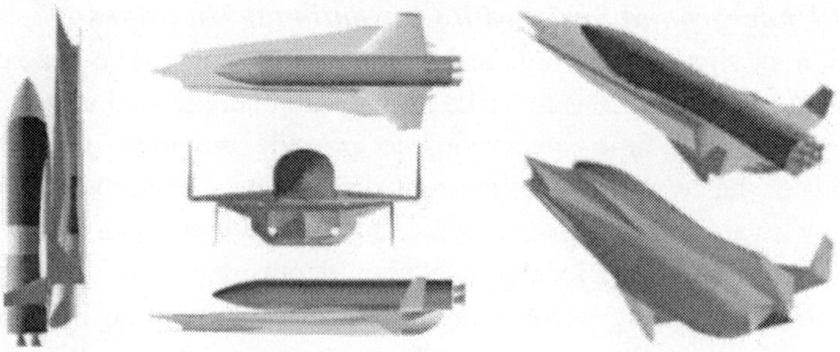

Figure 23. Airbreathing two-stage-to-orbit (TSTO) systems are based on a rocket-based combined-cycle upper stage in which scramjet propulsion eliminates the need to carry a large oxidizer mass, enabling a substantial reduction in the cost per unit mass brought to low Earth orbit.

Airbreathing combined-cycle two-stage-to-orbit launch systems have the ability to provide substantial reductions in space launch costs.

2.27. Rapidly Composable Small Satellites

Small satellites as in Figure 24 can be assembled, tested, launched, and functioning on orbit within days of operational requirement generation. These satellites are based on a plug-and-play/open-architecture approach using standards for self-describing components within a discoverable and auto-configuring system. Satellite control can be accomplished with standardized ground-stations. These architectures can be sufficiently flexible to allow spiral upgrade of individual components, cooperative control with other satellites, and the potential for on-orbit self-assembly. Sub-assemblies can include propulsion for attitude control, orbital maintenance and maneuver, laser and/or traditional communications, ISR, SSA, self-defense and other missions. Such plug-and-play architectures are based on a cross-domain enabling concept, since air and cyber systems can also be built from preexisting modular components using similar "open architecture" approaches.

Plug-and-play modular components allow lower-cost small satellites with "good enough" capabilities to be rapidly designed and assembled for operationally responsive space missions.

2.28. Fractionated/Distributed Space Systems

Constellations of smaller satellites operating cooperatively to perform a given mission set can provide greater survivability and greater ease of systematic upgrade. Architectures based on fractionation involve several functionally different satellite elements that cooperate to act as a single system, while distributed architectures involve many copies of identical elements that operate coherently to produce greater capability than each element could individually. The individual satellite elements make use of on-board data processing to minimize inter-constellation communications to achieve a highly secure and jam-resistant architecture. Such systems may utilize plug-and-play/open-architectures and allow seamless upgrade of the constellation's overall capability by adding or substituting new individual, small satellites.

Fractionated or distributed space systems can potentially provide greater on-orbit capability at lower cost than is achievable with traditional approaches, while allowing rapid reconstitution in the event that elements in the constellation are lost.

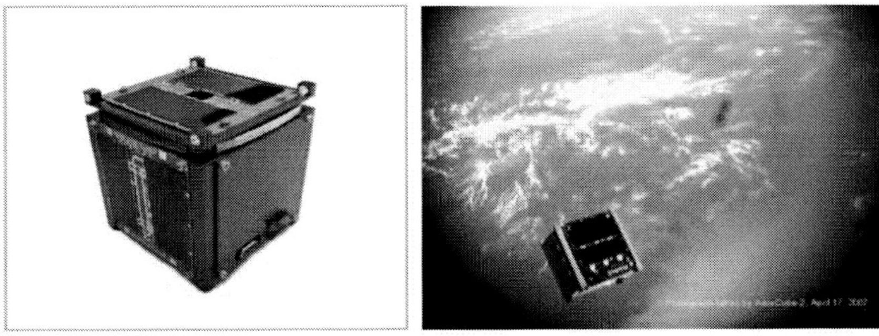

Figure 24. Small satellites composed from standardized modular elements can provide rapid on-orbit capabilities with "good enough" functionality. Fractionated or distributed space systems composed from such small satellites can significantly increase their functionality and survivability.

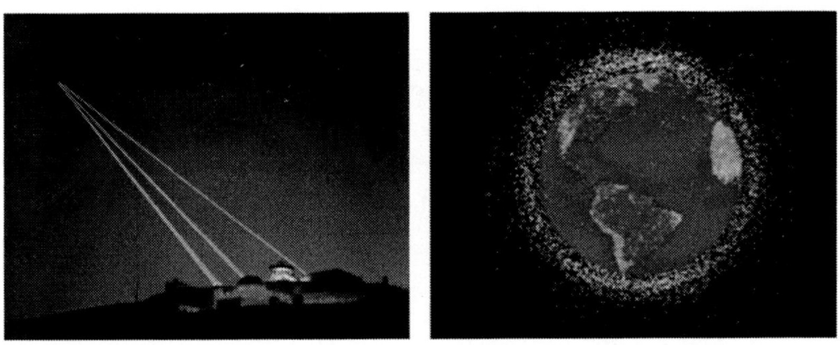

Figure 25. Improved space surveillance capability will consist of ground-based and space-based assets that can provide birth-to-death tracking and characterization of all spaceborne systems. Effective data fusion from multiple sources provides significant leverage over single-source observations.

2.29. Persistent Space Situational Awareness

Current ground-based radars and telescopes and space-based space surveillance assets that comprise the space surveillance network can be augmented to provide birth-to-death detection, tracking and characterization of every object in orbit from traditional large satellites to picosatellites and orbital debris at LEO, MEO, and GEO altitudes. As suggested in Figure 25, this can be done through a combination of new ground-based and space-based assets with appropriate fusion of data from other satellites, Aegis ships, and other sources into an integrated database. Current plans are to augment the radars and optical telescopes in the current space surveillance network with a 3.5-m Space Surveillance Telescope (SST), the Space Based Space Surveillance (SBSS) system, and the S-band Space Fence system. These will address detection and tracking, but determining the contents of a satellite or its potential capabilities and intent will be extremely difficult. Bringing together data from active and passive RF and EO/IR sources can provide a true "SSA network" with capabilities far beyond those of its individual elements. In principle, all satellites in orbit can provide information that when fused and analyzed provides a far more complete picture, including space weather effects to allow discrimination of hostile actions from natural causes.

A substantially greater space situational awareness capability will be essential to protecting space assets and allowing continued effective use of space.

2.30. Improved Orbital Conjunction Prediction

Development of a physics-based, near-real-time predictive model of near-earth space environment interactions with spacecraft will allow highly-accurate, actionable predictions of space object conjunctions. This model would utilize space weather data from ground- and space-based space environment sensors, space situational awareness sensor data of object locations/trajectories, and substantially improved astrodynamic calculations using improved satellite drag models. The model will provide confidence levels in conjunction predictions sufficient to justify orbital maneuvers to reduce risks of collisions.

Development of such an improved conjunction prediction capability will be essential for protecting space assets from collisions and from resulting increases in orbital debris that would endanger critical Air Force space assets.

3. Key Potential Capability Areas

The set of 30 potential capability areas (PCAs) are within reach as new Air Force capabilities by the 2030 target date of "Technology Horizons", and are matched to key needs of the strategic environment of this period. Among these PCAs, the following 12 are identified below as being of highest priority:

- PCA1: Inherently Intrusion-Resilient Cyber Systems
- PCA2: Automated Cyber Vulnerability Assessments
- PCA4: Augmentation of Human Performance
- PCA6: Adaptive Flexibly-Autonomous Systems
- PCA8: Dominant Spectrum Warfare Operations
- PCA9: PNT in GPS-Denied Environments
- PCA12: Processing-Enabled Intelligent ISR Sensors
- PCA15: Fractionated Survivable Remote-Piloted Systems
- PCA19: Next-Generation High-Efficiency Turbine Engines
- PCA24: Directed Energy for Tactical Strike/Defense
- PCA27: Rapidly Composable Small Satellites
- PCA29: Persistent Space Situational Awareness

4. From Potential Capability Areas to Key Technology Areas

The next Section lists the key technology areas that can enable each of the 30 PCAs identified here. In so doing, it identifies the S&T that need to be pursued over the next decade to allow these new Air Force capabilities to be developed. By correlating the resulting lists of key technology areas across all 30 PCAs, or across the top 12 PCAs listed above, it is possible to determine the S&T investments that can make the greatest contributions to Air Force capabilities for the 2010-2030 strategic environment.

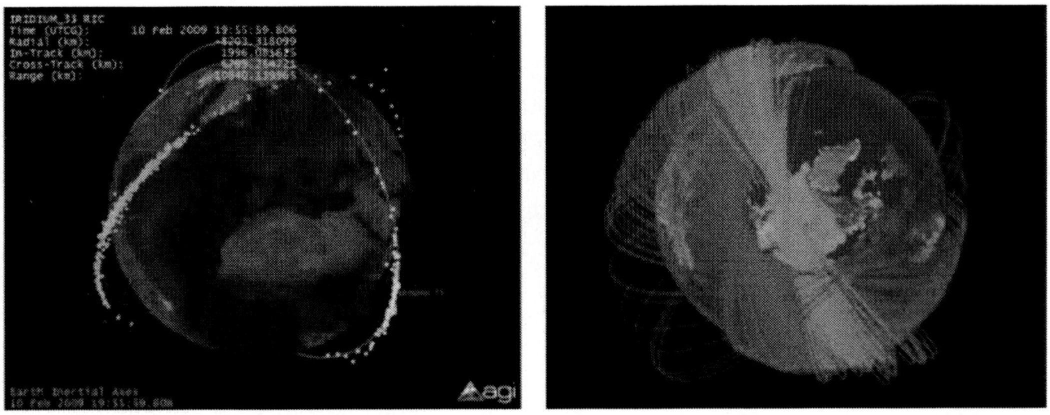

Figure 26. Beyond improved space situational awareness data, improved physics-level drag models for LEO and HEO objects are needed to enable orbital conjunction predictions with resulting uncertainty ellipses that are small enough to justify on-orbit avoidance maneuvers.

SECTION 6. KEY TECHNOLOGY AREAS DURING 2010-2030

This Section identifies a set of key technology areas that are among the most essential Air Force S&T to be pursued over the next decade, to enable technology-derived capabilities that can respond to the strategic environment, budget environment, and technology environment.

Each of the potential capability areas noted in the previous section is predicated on a set of underlying technologies that enable materiel or non-materiel means for achieving capabilities within that area. As indicted in Figure 6, the set of individual technology research efforts associated with these capabilities may be quite large, however these individual technologies can be sensibly aggregated into broader "technology areas".

Individual research efforts within any given technology area may be exploring different ways of achieving the same technology objective, or may be exploring different technology objectives that are needed to enable complementary aspects of a given technology. It is not the objective of "Technology Horizons" to dictate which of these individual *lines* of research within a given technology area will likely prove to be the most productive. Rather, this effort is appropriately focused on identifying which technology *areas* are most essential for Air Force S&T to pursue to enable potential capability areas such as those in Section 5 that are aligned with the strategic environment, enduring realities, and overarching themes identified in Sections 2-4.

Focusing on such key technology areas and not on individual lines of research is the appropriate level of specificity for an S&T vision. Identifying key technology areas provides useful guidance in making difficult choices among the limited set of S&T investments that can be pursued in the likely budget environment over the next two decades, yet it gives AFRL leadership the freedom to evaluate and select the specific individual lines of research that should be pursued.

1. Key Technology Areas Supporting Potential Capability Areas

For each of the potential capability areas identified in Section 5, a large number of supporting technologies are needed to allow systems of that type to be developed or to contribute to the increased effectiveness of such systems. It has not been an objective of "Technology Horizons" to produce an exhaustive list of all such supporting technologies. Indeed by being comprehensive such a list would be of little or no use in providing guidance for Air Force S&T decisions.

Instead this effort has sought to identify the more limited set of key technology areas that not only support these potential capability areas but also are broadly essential for enabling the overarching themes noted in Figure 5. The resulting "key technology areas" (KTAs) identified below for each of the "potential capability areas" (PCAs) should be understood in that way.

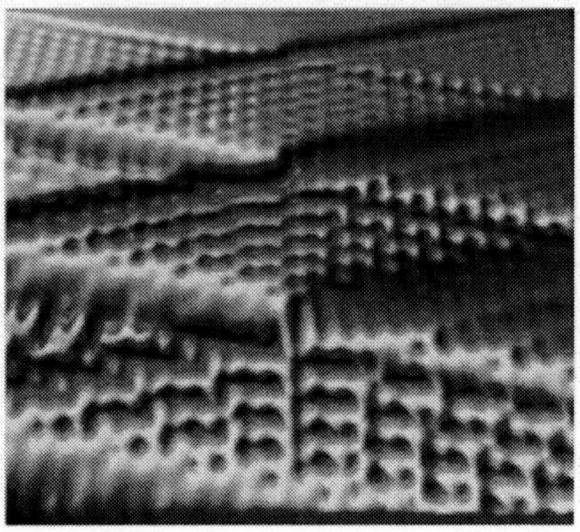

Basic research is providing the foundation on which a wide range of future capabilities can be based, such as superconducting artificial atom structures *(left)* visualized via amplitude spectroscopy, and metamaterials with negative refractive index that bend electromagnetic waves backwards *(right)*.

They do not define the entirety of S&T that the Air Force needs, but they are among the most essential components of the S&T that the Air Force will need for the strategic environment, enduring realities, and overarching themes that define the 2010-2030 time frame and beyond.

A set of 110 individual technology areas is identified here as being among the most essential S&T for Air Force to pursue over the next decade. Each of these "key technology areas" is mapped into the "potential capability areas" from Section 5 and the "overarching themes" from Section 4.

PCA1: Inherently Intrusion-Resilient Cyber Systems

- Ad hoc networks
- Virtual machine architectures
- Agile hypervisors
- Polymorphic networks
- Autonomous systems
- Autonomous reasoning
- Resilient autonomy
- Collaborative/cooperative control

- Agile networks
- Pseudorandom network recomposition
- Laser communications
- Secure RF links
- Frequency-agile RF systems
- Spectral mutability
- Dynamic spectrum access
- Quantum key distribution
- Complex adaptive distributed networks
- Complex adaptive systems
- Complex system dynamics
- V&V for complex adaptive systems
- Decision support tools
- Automated software generation
- Distributed sensing networks
- Sensor data fusion
- Signal identification and recognition
- Cyber offense
- Cyber defense
- Cyber resilience
- Advanced computing architectures
- Complex environment visualization
- Massive analytics
- Automated reasoning and learning

PCA2: Automated Cyber Vulnerability Assessments and Reactions

- Ad hoc networks
- Polymorphic networks
- Agile networks
- Complex environment visualization
- Complex adaptive distributed networks
- Frequency-agile RF systems
- Spectral mutability
- Dynamic spectrum access
- Quantum key distribution
- Complex adaptive systems
- Complex system dynamics
- V&V for complex adaptive systems
- Autonomous systems
- Autonomous reasoning and learning
- Resilient autonomy
- Collaborative/cooperative control
- Distributed sensing networks
- Decision support tools
- Information fusion and understanding
- Cyber defense
- Cyber resilience
- Human behavior modeling
- Cultural behavior modeling
- Social network modeling
- Complex environment visualization
- Massive analytics
- Human-machine interfaces

PCA3: Decision-Quality Prediction of Behavior

- Complex adaptive systems
- Complex system dynamics
- V&V for complex adaptive systems
- Autonomous systems
- Autonomous reasoning and learning
- Resilient autonomy
- Collaborative/cooperative control
- Decision support tools
- Distributed sensing networks
- Integrated sensing and processing
- Signal identification and recognition
- Information fusion and understanding
- Biological signatures
- Human behavior modeling
- Cultural behavior modeling
- Social network modeling
- Behavior prediction and anticipation
- Influence measures
- Cognitive modeling
- Human-machine interfaces

PCA4: Augmentation of Human Performance

- Chip-scale atomic clocks
- Ad hoc networks
- Virtual machine architectures
- Agile networks
- Complex adaptive distributed networks
- Complex adaptive systems
- Complex system dynamics
- V&V for complex adaptive systems
- Validation support to simulations
- Autonomous systems
- Autonomous reasoning and learning
- Resilient autonomy
- Collaborative/cooperative control
- Social network modeling
- Behavior prediction and anticipation
- Influence measures
- Cognitive modeling
- Complex environment visualization
- Autonomous mission planning
- Health monitoring and prognosis
- Decision support tools
- Automated software generation
- Distributed sensing networks
- Integrated sensing and processing
- Sensor-based processing
- Signal identification and recognition
- Information fusion and understanding
- Advanced computing architectures
- Biological signatures
- Human behavior modeling
- Cultural behavior modeling
- Massive analytics
- Cognitive performance augmentation
- Physical performance augmentation
- Human-machine interfaces

PCA5: Advanced Constructive Discovery & Training Environments

- Ad hoc networks
- Virtual machine architectures
- Agile networks
- Complex adaptive distributed networks
- Complex adaptive systems
- Complex system dynamics
- V&V for complex adaptive systems
- Multi-scale simulation technologies
- Coupled multi-physics simulations
- Validation support to simulations
- Autonomous systems
- Autonomous reasoning and learning
- Resilient autonomy
- Collaborative/cooperative control
- Autonomous mission planning
- Decision support tools
- Automated software generation
- Signal identification and recognition
- Information fusion and understanding
- Cyber resilience
- Advanced computing architectures
- Human behavior modeling
- Cultural behavior modeling
- Social network modeling
- Behavior prediction and anticipation
- Influence measures
- Cognitive modeling
- Complex environment visualization
- Massive analytics
- Human-machine interfaces

PCA6: Trusted, Adaptive, Flexibly-Autonomous Systems

- Complex adaptive systems
- Complex system dynamics
- V&V for complex adaptive systems
- Cold-atom INS
- Health monitoring and prognosis
- Decision support tools
- Automated software generation
- Distributed sensing networks

- Chip-scale atomic clocks
- Ad hoc networks
- Polymorphic networks
- Virtual machine architectures
- Agile networks
- Complex adaptive distributed networks
- Multi-scale simulation technologies
- Validation support to simulations
- Autonomous systems
- Autonomous reasoning and learning
- Resilient autonomy
- Collaborative/cooperative control
- Autonomous mission planning
- Embedded diagnostics
- Integrated sensing and processing
- Sensor-based processing
- Signal identification and recognition
- Information fusion and understanding
- Cyber defense
- Cyber resilience
- Advanced computing architectures
- Human behavior modeling
- Cultural behavior modeling
- Social network modeling
- Behavior prediction and anticipation
- Massive analytics
- Human-machine interfaces

PCA7: Frequency-Agile Spectrum Utilization

- Chip-scale atomic clocks
- Ad hoc networks
- Polymorphic networks
- Agile networks
- Pseudorandom network recomposition
- Complex adaptive distributed networks
- Laser communications
- Short-range secure RF communications
- Frequency-agile RF systems
- Spectral mutability
- Dynamic spectrum access
- Quantum key distribution
- Complex adaptive systems
- Complex system dynamics
- V&V for complex adaptive systems
- Autonomous systems
- Autonomous reasoning and learning
- Resilient autonomy
- Collaborative/cooperative control
- Autonomous mission planning
- Decision support tools
- Advanced RF apertures
- Secure RF links
- RF electronic warfare
- Distributed sensing networks
- Integrated sensing and processing
- Signal identification and recognition
- Information fusion and understanding
- Cyber resilience
- Advanced computing architectures
- Complex environment visualization
- Massive analytics
- Human-machine interfaces
- Optical and infrared materials
- RF and electronic materials
- Metamaterials

PCA8: Dominant Spectrum Warfare Operations

- Cold-atom INS
- Chip-scale atomic clocks
- Ad hoc networks
- Polymorphic networks
- Virtual machine architectures
- Agile networks
- Modular small-sat components
- High-power microwaves
- Lightweight multi-functional structures
- Autonomous systems
- Autonomous reasoning and learning
- Resilient autonomy
- Collaborative/cooperative control
- Passive radar

- Distributed small-sat architectures
- Fractionated small-sat architectures
- Laser communications
- Short-range secure RF communications
- Frequency-agile RF systems
- Spectral mutability
- Dynamic spectrum access
- Quantum key distribution
- Complex adaptive systems
- Complex system dynamics
- V&V for complex adaptive systems
- Directed energy effects
- Directed energy protection
- Optical and infrared materials
- RF and electronic materials
- Advanced RF apertures
- Secure RF links
- RF electronic warfare
- EO/IR sensing
- IR signature suppression
- Distributed sensing networks
- Integrated sensing and processing
- Sensor-based processing
- Signal identification and recognition
- Information fusion and understanding
- Cyber offense
- Cyber defense
- Cyber resilience
- Metamaterials
- Nanomaterials

PCA9: Precision Navigation/Timing in GPS-Denied Environments

- Cold-atom INS
- Chip-scale atomic clocks
- Ad hoc networks
- Polymorphic networks
- Agile networks
- Laser communications
- Short-range secure RF communications
- Frequency-agile RF systems
- Spectral mutability
- Quantum key distribution
- Complex adaptive systems
- Complex system dynamics
- V&V for complex adaptive systems
- Autonomous systems
- Autonomous reasoning and learning
- Resilient autonomy
- Collaborative/cooperative control
- Advanced RF apertures
- Secure RF links

PCA10: Next-Generation High-Bandwidth Secure Communications

- Cold-atom INS
- Chip-scale atomic clocks
- Ad hoc networks
- Laser communications
- Short-range secure RF communications
- Frequency-agile RF systems
- Spectral mutability
- Dynamic spectrum access
- Quantum key distribution
- Solid-state lasers
- Fiber lasers
- Semiconductor lasers
- EO/IR sensing
- Optical and infrared materials
- RF and electronic materials
- Metamaterials

PCA11: Persistent Near-Space Communications Relays

- Advanced aerodynamic configurations
- Aerodynamic test and evaluation
- Multi-scale simulation technologies
- Coupled multi-physics simulations

- Cold-atom INS
- Chip-scale atomic clocks
- Ad hoc networks
- Polymorphic networks
- Agile networks
- Laser communications
- Short-range secure RF communications
- Frequency-agile RF systems
- Spectral mutability
- Dynamic spectrum access
- Directed energy protection
- Lightweight multi-functional structures
- Advanced composite fabrication
- Structural modeling and simulation
- Metamaterials
- Self-healing materials
- Nanomaterials
- Material-specific manufacturing
- Power generation and energy storage
- Validation support to simulations
- Embedded diagnostics
- Health monitoring and prognosis
- High-altitude airship
- Advanced RF apertures
- Secure RF links
- RF electronic warfare
- Cyber defense
- Cyber resilience
- High-altitude materials
- Lightweight materials
- Advanced composites
- Composites sustainment
- RF and electronic materials
- High-temperature electronics
- Radiation hardened electronics
- System-level thermal management M&S
- Thermal management components
- High-OPR compressors

PCA12: Processing-Enabled Intelligent ISR Sensors

- Chip-scale atomic clocks
- Virtual machine architectures
- Complex system dynamics
- V&V for complex adaptive systems
- Quantum computing
- Decision support tools
- Advanced RF apertures
- EO/IR sensing
- Distributed sensing networks
- Integrated sensing and processing
- Sensor-based processing
- Signal identification and recognition
- Information fusion and understanding
- Cyber offense
- Cyber defense
- Cyber resilience
- Advanced computing architectures
- Biological signatures
- Massive analytics
- Autonomous reasoning and learning
- Optical and infrared materials
- RF and electronic materials
- Metamaterials
- Nanomaterials
- High-temperature electronics
- Radiation hardened electronics

PCA13: High-Altitude Long-Endurance ISR Airships

- Advanced aerodynamic configurations
- Aerodynamic test and evaluation
- Cold-atom INS
- Chip-scale atomic clocks
- Frequency-agile RF systems
- Spectral mutability
- Directed energy protection
- Advanced RF apertures
- Secure RF links
- RF electronic warfare
- EO/IR sensing
- High-altitude materials
- Lightweight materials
- Advanced composites

- Lightweight multi-functional structures
- Structural modeling and simulation
- Multi-scale simulation technologies
- Coupled multi-physics simulations
- Validation support to simulations
- Autonomous systems
- Autonomous mission planning
- Embedded diagnostics
- Health monitoring and prognosis
- High-altitude airship
- Metamaterials
- Nanomaterials
- Self-healing materials
- Material-specific manufacturing
- Power generation and energy storage
- High-temperature electronics
- Radiation hardened electronics
- System-level thermal management M&S
- Thermal management components

PCA14: Prompt Theater-Range ISR/Strike Systems

- Advanced aerodynamic configurations
- Cold-atom INS
- Chip-scale atomic clocks
- Advanced TPS materials
- Scramjet propulsion systems
- Frequency-agile RF systems
- Spectral mutability
- Directed energy protection
- Lightweight multi-functional structures
- Advanced composite fabrication
- Structural modeling and simulation
- Multi-scale simulation technologies
- Coupled multi-physics simulations
- Validation support to simulations
- Autonomous systems
- Autonomous mission planning
- Embedded diagnostics
- Aerodynamic test and evaluation
- Health monitoring and prognosis
- Advanced RF apertures
- Secure RF links
- EO/IR sensing
- IR signature suppression
- High-temperature materials
- High-altitude materials
- Lightweight materials
- Advanced composites
- Composites sustainment
- Material-specific manufacturing
- Hydrocarbon boost engine
- System-level thermal management M&S
- Thermal management components
- High-temperature fuel technologies
- Efficient bleedless inlets

PCA15: Fractionated, Survivable, Remotely-Piloted Systems

- Advanced aerodynamic configurations
- Aerodynamic test and evaluation
- Cold-atom INS
- Chip-scale atomic clocks
- Ad hoc networks
- Polymorphic networks
- Complex adaptive distributed networks
- Laser communications
- Short-range secure RF communications
- Frequency-agile RF systems
- Complex adaptive systems
- Collaborative/cooperative control
- Autonomous mission planning
- Embedded diagnostics
- Health monitoring and prognosis
- Advanced RF apertures
- Secure RF links
- RF electronic warfare
- EO/IR sensing
- IR signature suppression
- Cyber resilience
- Lightweight materials

- Complex system dynamics
- V&V for complex adaptive systems
- Directed energy effects
- Directed energy protection
- High-power microwaves
- Lightweight multi-functional structures
- Advanced composite fabrication
- Structural modeling and simulation
- Multi-scale simulation technologies
- Coupled multi-physics simulations
- Validation support to simulations
- Autonomous systems
- Autonomous reasoning and learning
- Resilient autonomy
- Advanced composites
- Composites sustainment
- Metamaterials
- Self-healing materials
- Material-specific manufacturing
- Power generation and energy storage
- System-level thermal management M&S
- Thermal management components
- Three-stream engine architectures
- High-OPR compressors
- Advanced / interturbine burners
- Efficient bleedless inlets
- Serpentine nozzles

PCA16: Direct Forward Air Delivery and Resupply

- Advanced aerodynamic configurations
- Aerodynamic test and evaluation
- Cold-atom INS
- Short-range secure RF communications
- Lightweight multi-functional structures
- Advanced composite fabrication
- Structural modeling and simulation
- Multi-scale simulation technologies
- Coupled multi-physics simulations
- Validation support to simulations
- Autonomous systems
- Resilient autonomy
- Embedded diagnostics
- Health monitoring and prognosis
- Decision support tools
- Advanced RF apertures
- Secure RF links
- Lightweight materials
- Advanced composites
- Composites sustainment
- Material-specific manufacturing

PCA17: Energy-Efficient Partially Buoyant Cargo Airlifters

- Advanced aerodynamic configurations
- Aerodynamic test and evaluation
- Directed energy effects
- Directed energy protection
- Lightweight multi-functional structures
- Advanced composite fabrication
- Structural modeling and simulation
- Multi-scale simulation technologies
- Coupled multi-physics simulations
- Validation support to simulations
- Autonomous systems
- Autonomous mission planning
- Embedded diagnostics
- Health monitoring and prognosis
- Advanced RF apertures
- Lightweight materials
- Advanced composites
- Composites sustainment
- Metamaterials
- Self-healing materials
- Nanomaterials
- Material-specific manufacturing
- Power generation and energy storage
- System-level thermal management M&S
- Thermal management components

PCA18: Fuel-Efficient Hybrid Wing-Body Aircraft

- Advanced aerodynamic configurations
- Aerodynamic test and evaluation
- Directed energy effects
- Directed energy protection
- Lightweight multi-functional structures
- Advanced composite fabrication
- Structural modeling and simulation
- Multi-scale simulation technologies
- Coupled multi-physics simulations
- Validation support to simulations
- Autonomous systems
- Embedded diagnostics
- Health monitoring and prognosis
- Advanced RF apertures
- IR signature suppression
- High-temperature materials
- Lightweight materials
- Advanced composites
- Composites sustainment
- Metamaterials
- Self-healing materials
- Nanomaterials
- Material-specific manufacturing
- System-level thermal management M&S
- Thermal management components
- Three-stream engine architectures

PCA19: Next-Generation High-Efficiency Turbine Engines

- Lightweight multi-functional structures
- Structural modeling and simulation
- Multi-scale simulation technologies
- Coupled multi-physics simulations
- Validation support to simulations
- Embedded diagnostics
- Health monitoring and prognosis
- High-temperature materials
- Lightweight materials
- Nanomaterials
- Nondestructive evaluation
- Material-specific manufacturing
- High-temperature electronics
- Alternate fuels
- System-level thermal management M&S
- Thermal management components
- Three-stream engine architectures
- High-temperature fuel technologies
- High-OPR compressors
- Engine component testing
- Advanced/interturbine burners
- Serpentine nozzles

PCA20: Embedded Diagnostic/Prognostic Subsystems

- Cold-atom INS
- Chip-scale atomic clocks
- Complex adaptive distributed networks
- Short-range secure RF communications
- Complex adaptive systems
- Complex system dynamics
- V&V for complex adaptive systems
- Fiber lasers
- Semiconductor lasers
- Lightweight multi-functional structures
- Multi-scale simulation technologies
- Coupled multi-physics simulations
- Validation support to simulations
- Decision support tools
- Advanced RF apertures
- Distributed sensing networks
- Integrated sensing and processing
- Signal identification and recognition
- Information fusion and understanding
- Cyber resilience
- High-temperature materials
- Lightweight materials
- Optical and infrared materials
- RF and electronic materials
- Metamaterials
- Nanomaterials

- Autonomous systems
- Autonomous reasoning and learning
- Resilient autonomy
- Embedded diagnostics
- Health monitoring and prognosis
- Nondestructive evaluation
- Material-specific manufacturing
- Power generation and energy storage
- High-temperature electronics

PCA21: Penetrating Persistent Long-Range Strike

- Advanced aerodynamic configurations
- Aerodynamic test and evaluation
- Ad hoc networks
- Polymorphic networks
- Complex adaptive distributed networks
- Laser communications
- Short-range secure RF communications
- Frequency-agile RF systems
- Structural modeling and simulation
- Multi-scale simulation technologies
- Coupled multi-physics simulations
- Validation support to simulations
- Autonomous systems
- Autonomous reasoning and learning
- Resilient autonomy
- Collaborative/cooperative control
- Autonomous mission planning
- Embedded diagnostics
- Health monitoring and prognosis
- Decision support tools
- Passive radar
- Advanced RF apertures
- Secure RF links
- RF electronic warfare
- EO/IR sensing
- IR signature suppression
- Integrated sensing and processing
- Information fusion and understanding
- Spectral mutability
- Complex adaptive systems
- Complex system dynamics
- V&V for complex adaptive systems
- Directed energy effects
- Directed energy protection
- Lightweight multi-functional structures
- Advanced composite fabrication
- Cyber offense
- Cyber resilience
- High-temperature materials
- Lightweight materials
- Advanced composites
- Composites sustainment
- Optical and infrared materials
- RF and electronic materials
- Metamaterials
- Self-healing materials
- Nanomaterials
- Material-specific manufacturing
- System-level thermal management M&S
- Thermal management components
- Three-stream engine architectures
- High-OPR compressors
- Advanced/interturbine burners
- Efficient bleedless inlets
- Serpentine nozzles

PCA22: High-Speed Penetrating Cruise Missile

- Advanced aerodynamic configurations
- Aerodynamic test and evaluation
- Cold-atom INS
- Chip-scale atomic clocks
- Advanced TPS materials
- Scramjet propulsion systems
- Advanced RF apertures
- Secure RF links
- RF electronic warfare
- Integrated sensing and processing
- Signal identification and recognition
- Information fusion and understanding

- Frequency-agile RF systems
- Complex adaptive systems
- Complex system dynamics
- V&V for complex adaptive systems
- Directed energy protection
- Lightweight multi-functional structures
- Advanced composite fabrication
- Structural modeling and simulation
- Multi-scale simulation technologies
- Coupled multi-physics simulations
- Validation support to simulations
- Autonomous systems
- Autonomous mission planning
- Embedded diagnostics
- Health monitoring and prognosis
- Cyber resilience
- High-temperature materials
- Lightweight materials
- Advanced composites
- Composites sustainment
- Metamaterials
- Self-healing materials
- Nanomaterials
- Material-specific manufacturing
- High-temperature electronics
- System-level thermal management M&S
- Thermal management components
- Engine component testing
- High-speed turbines
- Efficient bleedless inlets

PCA23: Hyperprecision Low-Collateral Damage Munitions

- Advanced aerodynamic configurations
- Aerodynamic test and evaluation
- Cold-atom INS
- Chip-scale atomic clocks
- Ad hoc networks
- Short-range secure RF communications
- Complex adaptive systems
- V&V for complex adaptive systems
- Lightweight multi-functional structures
- Advanced composite fabrication
- Structural modeling and simulation
- Multi-scale simulation technologies
- Coupled multi-physics simulations
- Validation support to simulations
- Autonomous systems
- Autonomous reasoning and learning
- Resilient autonomy
- Collaborative/cooperative control
- Autonomous mission planning
- Embedded diagnostics
- Advanced RF apertures
- Secure RF links
- EO/IR sensing
- Integrated sensing and processing
- Cyber resilience
- Lightweight materials
- Advanced composites
- Nanomaterials
- Material-specific manufacturing

PCA24: Directed Energy for Tactical Strike/Defense

- Advanced aerodynamic configurations
- Aerodynamic test and evaluation
- Complex adaptive systems
- Complex system dynamics
- V&V for complex adaptive systems
- Solid-state lasers
- Fiber lasers
- Semiconductor lasers
- Beam control
- Structural modeling and simulation
- Multi-scale simulation technologies
- Coupled multi-physics simulations
- Validation support to simulations
- Embedded diagnostics
- Health monitoring and prognosis
- High-temperature materials
- Lightweight materials
- Advanced composites

- Directed energy effects
- Directed energy protection
- High-power microwaves
- Lightweight multi-functional structures
- Advanced composite fabrication
- Power generation and energy storage
- High-temperature electronics
- System-level thermal management M&S
- Thermal management components
- Three-stream engine architectures

PCA25: Enhanced Underground Strike with Conventional Munitions

- Cold-atom INS
- Chip-scale atomic clocks
- Structural modeling and simulation
- Multi-scale simulation technologies
- Coupled multi-physics simulations
- Validation support to simulations
- Collaborative/cooperative control
- Autonomous mission planning

PCA26: Reusable Airbreathing Access-to-Space Launch

- Advanced aerodynamic configurations
- Advanced TPS materials
- Scramjet propulsion systems
- Complex adaptive systems
- Complex system dynamics
- V&V for complex adaptive system
- Lightweight multi-functional structures
- Advanced composite fabrication
- Structural modeling and simulation
- Multi-scale simulation technologies
- Coupled multi-physics simulations
- Validation support to simulations
- Embedded diagnostics
- Health monitoring and prognosis
- Advanced RF apertures
- Secure RF links
- Aerodynamic test and evaluation
- Cyber resilience
- High-temperature materials
- High-altitude materials
- Lightweight materials
- Advanced composites
- Composites sustainment
- Self-healing materials
- Nanomaterials
- Material-specific manufacturing
- Hydrocarbon boost engine
- High-temperature electronics
- Radiation hardened electronics
- System-level thermal management M&S
- Thermal management components
- Efficient bleedless inlets

PCA27: Rapidly Composable Small Satellites

- Complex adaptive distributed networks
- Modular small-sat components
- Fractionated small-sat architectures
- Laser communications
- Short-range secure RF communications
- Frequency-agile RF systems
- Directed energy effects
- Directed energy protection
- Space weather
- Orbital environment characterization
- Satellite drag modeling
- Secure RF links
- RF electronic warfare
- EO/IR sensing
- Integrated sensing and processing
- Cyber resilience
- High-altitude materials
- Lightweight materials
- Advanced composites
- Optical and infrared materials
- RF and electronic materials
- Metamaterials

- Space situational awareness
- Lightweight multi-functional structures
- Advanced composite fabrication
- Structural modeling and simulation
- Multi-scale simulation technologies
- Coupled multi-physics simulations
- Validation support to simulations
- Embedded diagnostics
- Health monitoring and prognosis
- Advanced RF apertures
- Nanomaterials
- Material-specific manufacturing
- Hydrocarbon boost engine
- Spacecraft propulsion
- Electric propulsion
- Power generation and energy storage
- High-temperature electronics
- Radiation hardened electronics
- System-level thermal management M&S
- Thermal management components

PCA28: Fractionated/Distributed Space Systems

- Ad hoc networks
- Polymorphic networks
- Modular small-sat components
- Distributed small-sat architectures
- Fractionated small-sat architectures
- Laser communications
- Short-range secure RF communications
- Frequency-agile RF systems
- Spectral mutability
- Complex adaptive systems
- Complex system dynamics
- V&V for complex adaptive systems
- Directed energy effects
- Directed energy protection
- High-power microwaves
- Space weather
- Orbital environment characterization
- Satellite drag modeling
- Space situational awareness
- Lightweight multi-functional structures
- Advanced composite fabrication
- Structural modeling and simulation
- Multi-scale simulation technologies
- Coupled multi-physics simulations
- Validation support to simulations
- Autonomous systems
- Collaborative/cooperative control
- Autonomous mission planning
- Agile networks
- Complex adaptive distributed networks
- Embedded diagnostics
- Health monitoring and prognosis
- Advanced RF apertures
- Secure RF links
- RF electronic warfare
- EO/IR sensing
- Distributed sensing networks
- Integrated sensing and processing
- Cyber resilience
- High-temperature materials
- High-altitude materials
- Lightweight materials
- Advanced composites
- Optical and infrared materials
- RF and electronic materials
- Metamaterials
- Self-healing materials
- Nanomaterials
- Hydrocarbon boost engine
- Spacecraft propulsion
- Electric propulsion
- Power generation and energy storage
- High-temperature electronics
- Radiation hardened electronics
- System-level thermal management M&S
- Thermal management components

PCA29: Persistent Space Situational Awareness

- Distributed small-sat architectures
- Fractionated small-sat architectures
- Multi-scale simulation technologies
- Coupled multi-physics simulations
- Validation support to simulations
- Decision support tools
- Advanced RF apertures
- EO/IR sensing
- Distributed sensing networks
- Integrated sensing and processing
- Information fusion and understanding
- Cyber resilience
- Complex environment visualization
- Massive analytics
- Optical and infrared materials
- RF and electronic materials

PCA30: Improved Orbital Conjunction Prediction

- Space weather
- Orbital environment characterization
- Satellite drag modeling
- Space situational awareness
- Multi-scale simulation technologies
- Advanced computing architectures
- Complex environment visualization
- Coupled multi-physics simulations
- Validation support to simulations
- Decision support tools
- Distributed sensing networks
- Information fusion and understanding
- Massive analytics

Not every key technology area listed under each potential capability area is necessarily relevant to every line of research that might support development of materiel or non-materiel means for achieving capabilities within that area. However all are potentially relevant to at least some key aspects of approaches that might be pursued under that capability area.

2. Alignment of Key Technology Areas with Overarching Themes

Section 4 identified a set of 12 broad overarching themes that help to define the most essential research efforts that Air Force S&T should pursue to prepare for the strategic environment and enduring realities of the 2010-2030 time frame. These overarching themes are aggregated into logical clusters below, and the key technology areas identified above are mapped into these to clarify how these themes can be used to guide S&T investment choices.

FROM … Platforms, Integrated, Preplanned, Long System Life to … Capabilities, Fractionated, Composable, Expendable

- Advanced aerodynamic configurations
- Aerodynamic test and evaluation
- Cold-atom INS
- Chip-scale atomic clocks
- Ad hoc networks
- Polymorphic networks
- Virtual machine architectures
- Fiber lasers
- Semiconductor lasers
- Beam control
- Directed energy effects
- Directed energy protection
- High-power microwaves
- Quantum computing

- Agile hypervisors
- Agile networks
- Pseudorandom network recomposition
- Complex adaptive distributed networks
- Modular small-sat components
- Distributed small-sat architectures
- Fractionated small-sat architectures
- Laser communications
- Short-range secure RF communications
- Frequency-agile RF systems
- Spectral mutability
- Dynamic spectrum access
- Quantum key distribution
- Complex adaptive systems
- Complex system dynamics
- V&V for complex adaptive systems
- Solid-state lasers
- Decision support tools
- Automated software generation
- Advanced RF apertures
- Secure RF links
- RF electronic warfare
- EO/IR sensing
- Distributed sensing networks
- Integrated sensing and processing
- Sensor-based processing
- Signal identification and recognition
- Information fusion and understanding
- Cyber resilience
- Advanced computing architectures
- Human behavior modeling
- Cultural behavior modeling
- Social network modeling
- Behavior prediction and anticipation
- Influence measures
- Cognitive modeling
- Complex environment visualization
- Massive analytics
- Space weather
- Orbital environment characterization
- Satellite drag modeling
- Space situational awareness
- Lightweight multi-functional structures
- Advanced composite fabrication
- Structural modeling and simulation
- Multi-scale simulation technologies
- Coupled multi-physics simulations
- Validation support to simulations
- Autonomous systems
- Autonomous reasoning
- Resilient autonomy
- Collaborative/cooperative control
- Autonomous mission planning
- Embedded diagnostics
- Health monitoring and prognosis
- Automated reasoning and learning
- Cognitive performance augmentation
- Physical performance augmentation
- Human-machine interfaces
- Lightweight materials
- Advanced composites
- Composites sustainment
- Optical and infrared materials
- RF and electronic materials
- Metamaterials
- Self-healing materials
- Nanomaterials
- Nondestructive evaluation
- Material-specific manufacturing
- Spacecraft propulsion
- Electric propulsion
- Power generation and energy storage
- High-temperature electronics
- Radiation hardened electronics
- System-level thermal management M&S
- Thermal management components

FROM … Cyber Defense, Permissive, Operations, Single-Domain to … Cyber Resilience, Contested, Dissuasion, Cross-Domain

- Cold-atom INS
- Chip-scale atomic clocks
- Ad hoc networks
- Complex system dynamics
- V&V for complex adaptive systems
- Directed energy effects

- Polymorphic networks
- Virtual machine architectures
- Agile hypervisors
- Agile networks
- Pseudorandom network recomposition
- Complex adaptive distributed networks
- Modular small-sat components
- Distributed small-sat architectures
- Fractionated small-sat architectures
- Laser communications
- Short-range secure RF communications
- Frequency-agile RF systems
- Spectral mutability
- Dynamic spectrum access
- Quantum key distribution
- Complex adaptive systems
- Health monitoring and prognosis
- Decision support tools
- Automated software generation
- Passive radar
- Advanced RF apertures
- Secure RF links
- RF electronic warfare
- EO/IR sensing
- IR signature suppression
- Distributed sensing networks
- Integrated sensing and processing
- Sensor-based processing
- Signal identification and recognition
- Information fusion and understanding
- Cyber offense
- Cyber resilience
- Advanced computing architectures
- Biological signatures
- Human behavior modeling
- Cultural behavior modeling
- Social network modeling
- Behavior prediction and anticipation
- Influence measures
- Directed energy protection
- High-power microwaves
- Quantum computing
- Space weather
- Orbital environment characterization
- Satellite drag modeling
- Space situational awareness
- Multi-scale simulation technologies
- Coupled multi-physics simulations
- Validation support to simulations
- Autonomous systems
- Autonomous reasoning
- Resilient autonomy
- Collaborative/cooperative control
- Autonomous mission planning
- Embedded diagnostics
- Complex environment visualization
- Massive analytics
- Automated reasoning and learning
- Cognitive performance augmentation
- Physical performance augmentation
- Human-machine interfaces
- High-altitude materials
- Lightweight materials
- Advanced composites
- Composites sustainment
- Optical and infrared materials
- RF and electronic materials
- Metamaterials
- Self-healing materials
- Nanomaterials
- Spacecraft propulsion
- Electric propulsion
- Power generation and energy storage
- High-temperature electronics
- Radiation hardened electronics
- Alternate fuels
- System-level thermal management M&S
- Thermal management components

FROM ... Sensors, Control, Single-Domain to ... Information, Autonomy, Cross-Domain

- Cold-atom INS
- Chip-scale atomic clocks
- Quantum key distribution
- Complex adaptive systems

- Ad hoc networks
- Polymorphic networks
- Virtual machine architectures
- Agile hypervisors
- Agile networks
- Pseudorandom network recomposition
- Complex adaptive distributed networks
- Modular small-sat components
- Distributed small-sat architectures
- Fractionated small-sat architectures
- Laser communications
- Short-range secure RF communications
- Frequency-agile RF systems
- Spectral mutability
- Dynamic spectrum access
- Collaborative/cooperative control
- Autonomous mission planning
- Embedded diagnostics
- Health monitoring and prognosis
- Decision support tools
- Automated software generation
- Secure RF links
- RF electronic warfare
- Distributed sensing networks
- Integrated sensing and processing
- Sensor-based processing
- Signal identification and recognition
- Information fusion and understanding
- Cyber resilience
- Advanced computing architectures
- Biological signatures
- Human behavior modeling
- Complex system dynamics
- V&V for complex adaptive systems
- Directed energy protection
- High-power microwaves
- Quantum computing
- Space weather
- Orbital environment characterization
- Satellite drag modeling
- Space situational awareness
- Multi-scale simulation technologies
- Coupled multi-physics simulations
- Validation support to simulations
- Autonomous systems
- Autonomous reasoning
- Resilient autonomy
- Cultural behavior modeling
- Social network modeling
- Behavior prediction and anticipation
- Influence measures
- Cognitive modeling
- Complex environment visualization
- Massive analytics
- Automated reasoning and learning
- Cognitive performance augmentation
- Human-machine interfaces
- Metamaterials
- Self-healing materials
- Nanomaterials
- Power generation and energy storage
- High-temperature electronics
- Radiation hardened electronics

FROM … Manned, Control, Long System Life to … Remote-Piloted, Autonomy, Expendable

- Advanced aerodynamic configurations
- Aerodynamic test and evaluation
- Cold-atom INS
- Chip-scale atomic clocks
- Ad hoc networks
- Polymorphic networks
- Virtual machine architectures
- Agile hypervisors
- Agile networks
- Lightweight multi-functional structures
- Advanced composite fabrication
- Structural modeling and simulation
- Multi-scale simulation technologies
- Coupled multi-physics simulations
- Validation support to simulations
- Autonomous systems
- Autonomous reasoning
- Resilient autonomy

- Pseudorandom network recomposition
- Complex adaptive distributed networks
- Modular small-sat components
- Frequency-agile RF systems
- Spectral mutability
- Dynamic spectrum access
- Complex adaptive systems
- Complex system dynamics
- V&V for complex adaptive systems
- Semiconductor lasers
- Directed energy effects
- Directed energy protection
- High-power microwaves
- Quantum computing
- Integrated sensing and processing
- Sensor-based processing
- Signal identification and recognition
- Information fusion and understanding
- Cyber resilience
- Advanced computing architectures
- Human behavior modeling
- Cultural behavior modeling
- Social network modeling
- Behavior prediction and anticipation
- Cognitive modeling
- Complex environment visualization
- Massive analytics
- Automated reasoning and learning
- Human-machine interfaces
- High-altitude materials
- Lightweight materials
- Advanced composites
- Collaborative/cooperative control
- Autonomous mission planning
- Embedded diagnostics
- Health monitoring and prognosis
- Decision support tools
- Automated software generation
- High-altitude airship
- Passive radar
- Advanced RF apertures
- Secure RF links
- RF electronic warfare
- EO/IR sensing
- IR signature suppression
- Distributed sensing networks
- Composites sustainment
- Optical and infrared materials
- RF and electronic materials
- Metamaterials
- Self-healing materials
- Nanomaterials
- Nondestructive evaluation
- Material-specific manufacturing
- Power generation and energy storage
- High-temperature electronics
- System-level thermal management M&S
- Thermal management components
- High-OPR compressors
- Engine component testing
- Advanced / interturbine burners
- Efficient bleedless inlets
- Serpentine nozzles
- High-speed turbines

FROM … Fixed, Single-Domain, Permissive to … Agile, Cross-Domain, Contested

- Advanced aerodynamic configurations
- Aerodynamic test and evaluation
- Cold-atom INS
- Chip-scale atomic clocks
- Ad hoc networks
- Polymorphic networks
- Virtual machine architectures
- Agile hypervisors
- Agile networks
- Solid-state lasers
- Fiber lasers
- Semiconductor lasers
- Beam control
- Directed energy effects
- Directed energy protection
- High-power microwaves
- Quantum computing
- Space weather

- Pseudorandom network recomposition
- Complex adaptive distributed networks
- Distributed small-sat architectures
- Fractionated small-sat architectures
- Laser communications
- Short-range secure RF communications
- Frequency-agile RF systems
- Spectral mutability
- Dynamic spectrum access
- Quantum key distribution
- Complex adaptive systems
- Complex system dynamics
- V&V for complex adaptive systems
- Passive radar
- Advanced RF apertures
- Secure RF links
- RF electronic warfare
- EO/IR sensing
- IR signature suppression
- Distributed sensing networks
- Integrated sensing and processing
- Sensor-based processing
- Signal identification and recognition
- Information fusion and understanding
- Cyber offense
- Cyber defense
- Cyber resilience
- Advanced computing architectures
- Human behavior modeling
- Cultural behavior modeling
- Social network modeling
- Behavior prediction and anticipation
- Influence measures
- Orbital environment characterization
- Satellite drag modeling
- Space situational awareness
- Lightweight multi-functional structures
- Multi-scale simulation technologies
- Coupled multi-physics simulations
- Validation support to simulations
- Resilient autonomy
- Collaborative/cooperative control
- Autonomous mission planning
- Embedded diagnostics
- Health monitoring and prognosis
- Decision support tools
- Cognitive modeling
- Complex environment visualization
- Massive analytics
- Automated reasoning and learning
- Cognitive performance augmentation
- Physical performance augmentation
- Human-machine interfaces
- Lightweight materials
- Optical and infrared materials
- RF and electronic materials
- Metamaterials
- Self-healing materials
- Nanomaterials
- Nondestructive evaluation
- Material-specific manufacturing
- Power generation and energy storage
- High-temperature electronics
- System-level thermal management M&S
- Thermal management components

3. Summary of Key Technology Areas

Among the 110 technology areas listed above, a smaller subset is common to many of these potential capability areas and overarching themes. This subset is especially important for enabling the shorter list of 12 top potential capability areas identified in the previous Section. This set represents key technology areas that are the most critical to advance over the next decade in order to enable Air Force capabilities that are matched to the 2030 environment.

- Autonomous systems
- Autonomous reasoning and learning
- Resilient autonomy
- Complex adaptive systems
- V&V for complex adaptive systems
- Collaborative/cooperative control
- Autonomous mission planning
- Cold-atom INS
- Chip-scale atomic clocks
- Ad hoc networks
- Polymorphic networks
- Agile networks
- Laser communications
- Frequency-agile RF systems
- Spectral mutability
- Dynamic spectrum access
- Quantum key distribution
- Multi-scale simulation technologies
- Coupled multi-physics simulations
- Embedded diagnostics
- Decision support tools
- Automated software generation
- Sensor-based processing
- Behavior prediction and anticipation
- Cognitive modeling
- Cognitive performance augmentation
- Human-machine interfaces

"If we are to achieve results never before accomplished, we must expect to employ methods never before attempted."

Sir Francis Bacon

The key technology areas identified above represent only a fraction of the research efforts currently being pursued in AFRL. However, they are identified here as being among the most essential S&T that must be pursued to enable Air Force capabilities that are matched to the 2010-2030 environment. Other supporting research efforts beyond these will also be needed to obtain a properly balanced and hedged portfolio, but these areas will be essential. These key technology areas will be needed to Air Force capabilities based on game-changing approaches such as flexible autonomy and human performance augmentation, which have been identified in "Technology Horizons" as central for maintaining Air Force technology superiority.

SECTION 7. GRAND CHALLENGES FOR AIR FORCE S&T DURING 2010-2030

This Section presents four focused technology challenges to help guide S&T efforts and drive innovative solutions in ways that support the enduring realities, overarching themes, and key technology areas identified in "Technology Horizons." Each leads to a demonstration of a substantial Air Force capability that requires significant S&T advances and integration across multiple technology areas.

To provide a set of identifiable focus problems that can help guide Air Force S&T over the next decade in the directions outlined by "Technology Horizons", the following "grand challenges" are presented. These are candidate problems designed to promote integration of S&T efforts across traditional technical domains in ways that can trigger creative means to solve them. Each is intended to lead to a major systems-level integrated technology demonstration. The problem statements are written to be sufficiently descriptive of a major technology-enabled "stretch capability" without prescribing the approaches by which the challenge may be met.

These are not meant to be competitions in the way that some DoD-related "grand challenges" have been. They are focus areas within which sets of individual key technology areas can be advanced and integrated at the whole-system level. The goal of each is to guide Air Force S&T in directions consistent with the enduring realities, overarching themes, and technology areas identified in "Technology Horizons", and enable dramatic advances over current capabilities in areas that will be essential for the Air Force over the next two decades and beyond.

By design, the scale and scope of these challenges vary significantly among them. Several may be solvable entirely by AFRL, while some may entail partnering with other organizations. Each is intended to stress the solution space and drive innovative development of key technologies. Beyond advancing the underlying technologies and their integration via these focused efforts, the demonstrations will help broaden Air Force-wide awareness of S&T advances in ways that could potentially lead to accelerated development of fieldable capabilities derived from them.

Challenge #1: Inherently Intrusion-Resilient Cyber Networks

Explore, develop, and demonstrate autonomous and scalable technologies that enable large, nonsecure networks to be made inherently and substantially more resilient to attacks entering through network or application layers, and to attacks that pass through these layers. Emphasis is on advancing technologies that enable network intrusion tolerance rather than traditional network defense, with the goal to achieve continued mission effectiveness under large-scale, diverse network attacks. Technologies may include, but are not limited to, virtualization, recomposability, IP hopping, pseudorandom switching, server self-cleansing, and other methods that provide substantial increases in resilience or tolerance to intrusions, are scalable to arbitrarily large networks, and do not require explicit network operator interventions. Applies overarching themes of agility, autonomy, composability, and resilience in cyber networks.

Demonstrate individual or combined benefits of resulting technologies in a representative, synthesized, large-scale, open network. This will include an extended period during which unrestricted parties worldwide are invited to achieve one or more of a set of defined goals that may include installation of malware or other implants, metatag alteration, data corruption, data exfiltration, distributed denial of service, and other representative predetermined objectives, in addition to provable accomplishment of any unspecified modifications in the network or interruptions of network functions.

Target date for demonstration: No later than 2015.

This challenge applies the overarching themes of agility, autonomy, composability, and resilience to cyber systems, and will result in methods for achieving substantially improved intrusion-resistance and intrusion-tolerance in Air Force cyber systems.

Challenge #2: Trusted Highly-Autonomous Decision-Making Systems

Explore, develop, and demonstrate technologies that enable current human-intensive functions to be replaced, in whole or in part, by more highly autonomous decision-making systems, and technologies that permit reliable verification and validation (V&V) to establish the needed trust in them. Emphasize decision-making systems requiring limited or no human intervention for current applications where substantial reductions in specialized manpower may be possible, and future applications involving inherent decision time scales far exceeding human capacity.

Technologies may include, but are not limited to, information fusion, cognitive architectures, robust statistical learning, search and optimization, automated reasoning, neural networks, complex system dynamics, and other approaches that will enable increasingly autonomous decision-making. In parallel, also develop generalized V&V methods for highly autonomous, adaptive, near-infinite state systems that can provide asymptotic inferential assessments of confidence levels for such systems over their state space. Emphasis in both areas should be on identifying broadly applicable principles, supporting theoretical constructs, practical methods and their algorithmic embodiments, and measures of effectiveness, rather than focusing primarily on individual application-specific instantiations.

Demonstrate effectiveness of the resulting technologies in a representative, synthesized, large-scale CAOC environment using recorded operational or operationally realistic data streams, with total or partial autonomous decision-making for functions such as battle management, deployment planning, and logistics management. Address resilience of decision-making systems under extremely high data stream capacities and exceptional events, including resilience to corrupted data or information. Quantify autonomous or semi-autonomous decision quality and compare with human operator decision-making results. Quantify levels of trust in decision-making systems involved, and compare with demonstration results.

Target date for demonstration: 2017.

Focused effort on this challenge will enable technologies that can support substantial manpower cost reductions, and extend robust improved decision-making capabilities to highly stressing future applications that may involve decision time scales beyond human capacity.

Challenge #3: Fractionated, Composable, Survivable, Autonomous Systems

Explore, develop, and demonstrate technologies that can enable future autonomous aircraft or spacecraft systems achieving greater multi-role capability across a broader range of missions at moderate cost, including increased survivability in contested environments. Emphasis is on composability via system architectures based on fractionation and redundancy. Involves advancing methods for collaborative control and adaptive autonomous mission planning, and verification and validation (V&V) of highly adaptable, autonomous control systems. Includes technology advances to enable autonomous coordinated flight operation of fractional elements using short-range, low-bandwidth, jam-resistant, secure communication links. May include RF frequency agility, burst transmissions, laser links, and other methods to achieve low probability of detection and maintain needed levels of link integrity. System-level optimization will determine degree of fractionation and define functionality of each type of fractional element.

Demonstrate underlying technologies in a fully integrated ground-based hardware-in-the-loop simulation of a complete fractionated system operating over of range of missions and denial environments. Incorporate representative fractional elements and level of redundancy needed for anti-access/area-denial operations. Include active links between all system elements for data exchange and C2 communications. Verify maintenance of link integrity over range of jamming levels and broader EW attacks. Show resilience of system functionality to various types and degrees of degradation in fractional element capabilities. Demonstrate mission resilience to cyber attacks and to loss of multiple fractional elements.

Target date for demonstration: No later than 2018.

This challenge helps focus development of technologies to enable composable systems of remote-piloted aircraft from sets of common, relatively low-cost, fractional elements that cooperate as a single system to provide flexible capabilities across a broader range of missions.

Challenge #4: Hyper-Precision Aerial Delivery in Difficult Environments

Explore, develop, and demonstrate technologies that enable single-pass, extremely precise, autonomously guided aerial delivery of equipment and supplies under GPS-denied conditions from altitudes representative of operations in mountainous and contested environments and winds representative of steep mountainous terrain. Emphasis on low-cost autonomous flight systems with control authority capable of reaching target point within specified impact limits under effects of large stochastic disturbances. Guidance technologies could include independent IMUs, back-referencing to delivery aircraft, or other methods. Technologies should address full range needed to enable complete air delivery system, including flight vehicle, guidance, terminal impact control, system health monitoring, and interfaces to airborne platform and ground units.

Demonstrate resulting technologies by consistently delivering payloads in excess of 2,000 lbs from 25kft altitude to within 25 meters of a designated point in steep terrain under wind conditions representative of mountainous areas. Integrated demonstration system should be deployable from current and future airlift systems, including remote-piloted platforms, in a single pass over the drop point. Impact metrics must meet standards for survivable delivery of equipment and supplies. System should be linked to GIG to report position, manifest, and system health information, but remain resilient to cyber attacks. Must be recoverable, and if possible be self-recoverable.

Target date for demonstration: 2018.

This challenge advances and integrates technologies in several key areas to demonstrate the potential of autonomous systems to meet extreme requirements of a key Air Force mission.

SECTION 8. SUMMARY OF "TECHNOLOGY HORIZONS" VISION

This Section presents a summary of the S&T vision from "Technology Horizons", and identifies the most critical focus areas for S&T investment during the next decade. Sustained research in these areas will provide the basis for Air Force capability dominance during 2010-2030 and beyond.

"Technology Horizons" has developed a realistically implementable vision of the most essential Air Force S&T that will be needed to meet the strategic, technological, and budget demands of 2010-2030. Before summarizing the major insights from this effort, the range of inputs that went into developing them and the key elements of the resulting S&T vision are summarized.

Broad Range of Inputs to "Technology Horizons"

The most important science and technology efforts to be pursued by the Air Force over the next decade and beyond as identified in this report are the result of an exceptionally broad range of inputs:

- They reflect inputs from working groups composed of individuals from the Air Force S&T community, intelligence community, MAJCOMs, Product Centers, FFRDCs, defense industry, and academia.
- They also reflect broad S&T-related perspectives from visits, discussions, and briefings with organizations across the Air Force, including Air Staff and Air Force Secretariat offices, MAJCOMs, Product Centers, Direct Reporting Units and Field Units.
- They further reflect additional inputs obtained from visits, discussions, and briefings with organizations representing the Department of Defense, federal agencies, FFRDCs, national laboratories, and industry.
- They additionally reflect operational perspectives from briefings, visits and discussions with Air Combat Command (ACC), Air Force Special Operations Command (AFSOC), Air Force Space Command (AFSPC), and Air Mobility Command (AMC).
- They also reflect technological, operational, and strategic perspectives in reports from the Air Force Scientific Advisory Board, Defense Science Board, Center for Strategic and Budgetary Assessments, and numerous other organizations.

Collectively, the broad range of perspectives that was considered in "Technology Horizons" ensures that the resulting S&T vision reflects key Air Force needs during 2010-2030 as well as the principal technology-enabled opportunities where properly focused S&T investments can best serve to meet those needs.

Elements of the S&T Vision from "Technology Horizons"

The role of any S&T vision is to help guide the right choices that must be made among a very large set of possible technology development efforts that could be pursued if resources were available. In the likely budget-constrained environment that the Air Force faces, such a guiding vision is essential for making these difficult choices. "Technology Horizons" thus provides an Air Force S&T vision for 2010-2030 that consists of the following elements:

Strategic Context

Principal strategic factors that will drive needed Air Force capabilities during this time, and factors relevant to the worldwide S&T arena that will impact the Air Force's ability to maintain superior technological capabilities.

Enduring Realities

Key drivers that will remain largely unchanged and that will act to constrain the ability of the Air Force to shape itself through S&T investments as it might otherwise be able to do in an unconstrained environment.

Overarching Themes

Specific overarching themes that will be central for meeting Air Force needs resulting from the strategic context and enduring realities; these themes form the essential foundation for the most important Air Force S&T during 2010-2030.

Potential Capability Areas

A set of technologically achievable capability areas that are well aligned with the overarching themes identified above, and that identify key enabling technologies that will be among the highest-value S&T investments during this time.

Key Technology Areas

Cross-cutting technology areas that will be among the most essential for enabling key potential capabilities such as those above, and will be among the most important areas to emphasize in Air Force S&T over the next decade and beyond.

Grand Challenges

A set of challenge problems to focus Air Force S&T over the next decade on many of the key enabling technology areas identified above, and to drive innovative approaches for solving the most important technical issues in these areas.

Implementation Plan and Recommendations

The proposed plan for implementing the S&T vision from "Technology Horizons", including actionable recommendations to vector Air Force S&T over the next decade together with corresponding primary and supporting organizations to implement these.

Essential Focus Areas for Air Force S&T Investment

A major insight to emerge from "Technology Horizons" is that Air Force science and technology over the next decade will need to focus as much on advancing technologies that enable reduced Air Force operating costs as on technologies that support more traditional development of new systems or capabilities. This includes technologies to reduce manpower costs, energy costs, and sustainment costs. Of these, manpower costs are by far the largest, yet research specifically directed at increasing manpower efficiencies or reducing manpower needs has to date received substantially less attention as an identifiable Air Force S&T focus area.

While research into increased energy efficiency, improved energy storage, and alternative energy sources must continue, as must research to enable reduced sustainment costs, a significant new focus must be placed on technologies to achieve reduced manpower needs and greater manpower efficiencies. Regarding the latter, "Technology Horizons" has identified two key areas in which substantial advances will be possible in the next decade through properly focused Air Force S&T investments: *(i)* increased use of autonomy and autonomous systems, and *(ii)* augmentation of human performance. Both offer enormous potential for achieving capability increases and operating cost savings via improved manpower efficiencies and reduced manpower needs.

1. Increased Use of Autonomy and Autonomous Systems

A key finding is the need, opportunity, and potential to dramatically advance technologies that can allow the Air Force to gain capability increases, manpower efficiencies, and cost reductions through far greater use of autonomous systems in essentially all aspects of Air Force operations. Flexibly autonomous systems can be applied far beyond the current remotely-piloted aircraft, operational flight programs, and other implementations in use today. Dramatically increased use of autonomy -- not only in the number of systems and processes to which autonomous control and reasoning can be applied but especially in the degree of autonomy that is reflected in these -- can offer the Air Force potentially enormous increases in its capabilities, and if implemented correctly can do so in ways that enable manpower efficiencies and cost reductions.

Beyond the efficiency increases and associated manpower cost reductions that are achievable through such greater use of highly adaptible and flexibly autonomous systems and processes, there are very significant time-domain operational advantages to be gained over adversaries who are limited to human planning and decision speeds. The increased ops tempo that is possible through greater use of autonomous systems itself represents a significant capability advantage.

Achieving these gains from greater use of autonomous systems will require development of entirely new methods for enabling "trust in autonomy" through verification and validation (V&V) of the near-infinite state systems that result from high levels of adaptibility and autonomy. In effect, the number of possible input states that such systems can be presented with is so large that not only is it impossible to test all of them directly, it is not even possible to test more than an insignificantly small fraction of them. Proper development of such highly adaptive autonomous systems is thus inherently unverifiable by today's methods, and as a result their operation in all but comparatively trivial applications is uncertifiable.

Already today it is technically possible to imagine and build systems having immense levels of autonomy, but it is the lack of suitable V&V methods that prevents errors in the development of such systems from being detected, and thereby prevents all but relatively low levels of autonomy from being certified for use. Potential adversaries, however, may be entirely willing to field systems that benefit from far higher levels of autonomy without imposing any need for certifiable V&V, and could gain potentially significant capability advantages over the Air Force by doing so. The ease with which such highly autonomous systems can be designed and built in comparison with the substantial burden required to develop V&V methods for certifying them creates an inherent asymmetry that favors adversaries. Countering this asymmetric advantage will require access to as-yet undeveloped methods for achieving certifiably reliable V&V.

Developing certifiable V&V methods for highly adaptive autonomous systems is one of the major challenges facing the entire field of control science, and one that may require the larger part of a decade or more to develop a fundamental understanding of the underlying theoretical principles and various ways that these could be applied. The Air Force, as one the greatest potential beneficiaries of more highly adaptive and autonomous systems, must be a leader in the development of the underlying science and technology principles for V&V.

2. Augmentation of Human Performance

A second key finding to emerge from "Technology Horizons" is that natural human capacities are becoming increasingly mismatched to the enormous data volumes, processing capabilities, and decision speeds that technologies either offer or demand. Although humans today remain more capable than machines for many tasks, by 2030 machine capabilities will have increased to the point that humans will have become the weakest component in a wide array of systems and processes. Humans and machines will need to become far more closely coupled, through improved human-machine interfaces and by direct augmentation of human performance.

The field of human performance augmentation has made considerable advances over the past two decades, and with appropriate focused research efforts over the next decade will permit significant practical instantiations of augmented human performance. Such augmentation may come from increased use of autonomous systems as noted above, from improved man-machine interfaces to couple humans more closely and more intuitively with automated systems, and even from direct augmentation of humans themselves. The latter includes neuropharmaceuticals or implants to improve memory, alertness, cognition, or visual/aural acuity, as well as screening of individuals for speciality codes based on brainwave patterns or genetic correlators, or even genetic modification itself. While such methods may be inherently distasteful to some, potential adversaries may be entirely willing to make use of them.

Developing acceptable ways of using science and technology to augment human performance will become increasingly essential for gaining the benefits that many technologies can bring. The current technical maturity of various approaches in this area varies widely, but significant steps to advance and develop early implementations are possible over the next decade. Performance augmentation provides a further means to increase human efficiencies, allowing reduced manpower needs for the same capabilities or increased capabilities with given manpower.

3. Technologies to Increase Freedom of Operations in Contested Environments

A further key theme from "Technology Horizons" is the need to focus a greater fraction of Air Force S&T investments on research to support increased freedom of operations in contested or denied environments. Three main research areas are found to be of particular importance in this connection:

Cyber Resilience

Research into technologies for increased cyber resilience differs in subtle but important ways from the current emphasis being placed on cyber defense. While defense is focused on preventing adversaries from entering cyber systems, resilience refers to technologies that inherently make cyber systems far more difficult for an adversary to exploit once entry is gained, or that allow cyber systems to operate more effectively even when under overt cyber attack. Cyber resilience thus supports "fighting through" approaches that seek to maintain mission assurance across the entire spectrum of cyber threat levels.

Technologies such as massive virtualization, agile hypervisors, and inherent polymorphism can enable cyber systems that are fundamentally more resilient to intrusions. They greatly complicate a cyber adversary's ability to plan and coordinate effective cyber attacks by reducing the time over which our networks remain essentially static. In so doing they may also cause a cyber intruder to leave behind greater forensic evidence for attribution. It is likely that potential adversaries will also recognize the value of resilience in their own cyber networks, though our current greater dependence on cyber systems makes our development of such technologies more imperative.

The role of cyber operations will continue to grow as one of the foundational components of conflict escalation control and warfare. Beyond the defensive benefits of inherently resilient cyber environments, such technologies can also provide new means for expressing changes in defensive posture and signaling levels of escalation during periods of tension. For instance, by progressively increasing the rate of network polymorphism as tensions increase, greater network performance can be traded off to gain increased network resilience in a way that is readily detectable by an adversary and that allows signaling an increased state of preparedness to survive cyber attacks.

Precision Navigation & Timing in GPS-Denied Environments

The second focus area needed to enable increased freedom of operations is research on technologies to augment or supplant current precision navigation and timing (PNT) in GPS-denied environments. These include chip-scale inertial measurement units and atomic clocks, as well as currently less mature "cold atom" inertial navigation systems and timing systems based on compact matter-wave interferometry approaches.

The latter make use of various phenomena associated with Bose-Einstein condensates that can be realized with magneto-optical traps and other approaches currently being explored. The resulting ultra-precise position and timing capabilities may allow GPS-like accuracies relative to an initial reference to be maintained for sufficiently long periods to offset much of the benefit to an adversary of local GPS denial. In addition to such ultra-precise PNT augmentation there is a need for research into improved terrain matching and other less accurate but robust approaches that can provide position information under broader GPS denial. The dependence of current Air Force systems on availability of PNT information makes research efforts to develop GPS surrogate technologies essential.

Electromagnetic Spectrum Warfare

The third research area to support freedom of operations in contested environments is in technologies for dominant electromagnetic spectrum warfare capabilities. These include various approaches for enabling greater spectral mutability to increase waveform diversity, including methods for pulse-to-pulse radar waveform encoding that can increase resilience to spoofing and resistance to signal injection. Dynamic spectrum access methods may also provide greater resilience to jamming and other modes of electronic attack, and can give resilience to lost spectrum as bands are transferred to commercial uses. New methods for electronic attack are also needed to offset increasing adversary use of advanced integrated air defenses. Research on ultrawideband RF aperture technologies will be needed to allow spectrum warfare capabilities to be cost-effectively integrated into platforms.

Additional High-Priority Technology Areas

"Technology Horizons" has further identified the following as being among key priority areas where S&T investment will be needed over the next decade to enable essential capabilities:

- ***Processing-enabled intelligent sensors:*** ISR sensors with backplane processing for data synthesis and fusion to permit cueing-level PED functions to be performed on the sensor itself, reducing bandwidth otherwise consumed in transferring large amounts of raw data to the ground and manpower consumed for cueing-level processing.
- ***Directed energy for tactical strike/defense:*** Laser and high power microwave systems with sufficient power and beam control for tactical strike, air vehicle defense, and active airbase defense applications.
- ***Persistent space situational awareness:*** Birth-to-death detection, tracking and characterization of objects in orbit from traditional large satellites to picosatellites and orbital debris at LEO, MEO, and GEO altitudes, using ground- and space-based assets with data fusion from other sources into an integrated database
- ***Rapidly composable small satellite systems:*** Satellites that can be assembled, tested, and launched within days of operational requirement, based on a plug-and-play/open-architecture approach using standards for self-describing components within a discoverable and auto-configuring system.
- ***Next-generation high-efficiency gas turbine systems:*** A wide range of technology advances that collectively can enable substantial improvements in turbine engine fuel efficiency needed for long-range strike and long-endurance ISR systems.

Several further key themes can be identified in the results from "Technology Horizons". These include technology-derived means for achieving substantial fuel cost savings, in part through nontraditional flight vehicles such as hybrid wing-body aircraft, high-altitude long-endurance airships, and partially buoyant cargo airlifters. While some such systems may not be fielded by the 2030 horizon date of this study, research to advance these technologies is appropriate to provide options when fuel prices become incompatible with the fuel efficiencies achievable by more conventional flight vehicles.

"*The future is here. It is just not widely distributed yet.*"

William Gibson

Science fiction author who coined the term "cyberspace" in 1984

SECTION 9. IMPLEMENTATION PLAN & RECOMMENDATIONS

An implementation plan is presented with specific actionable recommendations for vectoring S&T in directions consistent with findings from "Technology Horizons" to maximize the technology advantage of the U.S. Air Force over the next decade and beyond.

1. Recommendation #1: Communicate Results from "Technology Horizons"

The first step involves communicating the rationale, objectives, and process for this study and the key elements of the resulting S&T vision across the Air Force in order to build broad awareness, understanding, and support for its implementation.

1.1. Brief "Technology Horizons" to Key Air Force Organizations

"Technology Horizons" should be offered as a brief to all HAF offices, MAJCOMs and Product Centers, AFRL, and elsewhere as appropriate as a further means for disseminating its findings and seeking additional inputs.

1.2. Disseminate "Technology Horizons" Report

Volumes 1 and 2 should be disseminated widely across key Air Force organizations beyond the HAF offices, MAJCOM and Product Center leadership, and Air Force S&T leadership that have provided inputs to the study. This effort should include operational and supporting components as a way of developing broad awareness, understanding, and support for Air Force S&T goals over the next decade.

"Technology Horizons" should also be disseminated across DoD, including DDR&E and DARPA, the other Services including the Office of Naval Research (ONR), Naval Research Laboratory (NRL), Army Research Office (ARO), and U.S. Special Operations Command (USSOCOM), and to NASA and other Federal agencies, as well as FFRDCs, academia and industry organizations.

This may include briefings made to these organizations, industry groups, and others outlining the goals and S&T focus areas identified in "Technology Horizons". These organizations will want to coordinate their S&T efforts with those of the Air Force, and are likely to be partners in implementing some aspects of the Air Force S&T vision.

- Communicate "Technology Horizons" rationale, objectives, process, and key elements via briefings offered to all HAF offices, MAJCOMs, Product Centers, and AFRL.
- Build broad awareness, understanding, and support for the Air Force S&T vision from "Technology Horizons".
- Disseminate Volumes 1 and 2 of "Technology Horizons" across all relevant organizations beyond those that provided inputs to this effort.

Effectively communicating the Air Force S&T vision from "Technology Horizons" is the first step in developing broad awareness, understanding, support, and embrace for its objectives in order to enable its subsequent implementation steps.

2. Recommendation #2: Assess Alignment of S&T Portfolio with "Technology Horizons"

The second implementation step is an assessment by AFRL of the alignment of its current S&T portfolio with the broad research directions and technology focus areas outlined in this report. A goal of this is to determine the extent to which the current portfolio is aligned, and the fraction of the portfolio that should be aligned with these goals.

2.1. Determine Alignment of Current S&T Portfolio with "Technology Horizons"

Many of the technology focus areas where Air Force S&T should be focused to meet emerging Air Force needs are already being emphasized to various degrees in research currently being done in AFRL. In other areas there is insufficient or no work being done on the science and technology needed to enable these potential capability areas.

Alignment of the current Air Force S&T portfolio with "Technology Horizons" should be assessed.

2.2. Identify Fraction of Portfolio to be Aligned with "Technology Horizons"

As noted throughout this report, no effort such as "Technology Horizons" can usefully list all S&T that is important for the Air Force to pursue. It must therefore be recognized that the key technology areas identified here represent only a fraction of the overall research portfolio that should be pursued in AFRL. In addition to assessing the alignment of its current science and technology efforts with "Technology Horizons", the Air Force should identify the fraction of its total S&T portfolio that it believes should be aligned with the focus areas identified here.

- Assess alignment of AFRL's current S&T portfolio with the broad research directions and technology focus areas outlined in "Technology Horizons".
- Identify target fraction of the total Air Force S&T portfolio to be aligned with the research directions and technology focus areas identified in "Technology Horizons".

An appropriate fraction of Air Force S&T efforts should be consistent with key areas identified in "Technology Horizons" as being essential to meet Air Force needs during 2010-2030 and beyond.

3. Recommendation #3: Adjust S&T Portfolio Balance as Needed

The third step is the identification by AFRL of research efforts in its current S&T portfolio that require redirection or realignment with the vision in "Technology Horizons", as well as new research that must be started to support these goals. These changes will redirect focus and emphasis in key areas.

3.1. Identify Current Efforts Requiring Realignment or Redirection

In the portion of its research portfolio that should be aligned with "Technology Horizons", AFRL will use its existing corporate process to identify current research efforts that are not well aligned with this vision. For each of those efforts, AFRL will determine if the effort can be redirected or realigned appropriately, or if it should be terminated to accommodate a more constructively formulated effort that achieves the needed direction and emphasis.

3.2. Determine New S&T Efforts That must be Started

Also using the existing corporate process, AFRL will define new research efforts that enable the key goals identified in "Technology Horizons" to be effectively pursued. The resources needed to enable these new efforts may come from efforts identified above for realignment, redirection, or termination, and they may come from new resources dedicated specifically to enabling the vision in "Technology Horizons".

- Identify research efforts in current S&T portfolio that must be redirected or realigned with the research directions and technology focus areas in "Technology Horizons".
- Determine which of these efforts should be realigned, redirected, or terminated to accommodate new research efforts that achieve the needed direction and emphasis.
- Define new research efforts that will be started to allow broad research directions and technology areas identified in "Technology Horizons" to be effectively achieved.
- Implement changes in the AFRL S&T portfolio to initiate new research efforts identified above, and to realign, redirect, and terminate existing efforts identified above.

A mix of "buy, hold, and sell" positions will be needed among research efforts in the current Air Force S&T portfolio to achieve the S&T vision in "Technology Horizons".

4. Recommendation #4: Initiate Focused Research on "Grand Challenge" Problems

As the next step after adjusting the current S&T portfolio, AFRL should evaluate and refine a set of "grand challenge" problems similar to those identified here, and use these to focus major technology development efforts in key areas identified in "Technology Horizons". The specific challenge problems to be undertaken will be defined by the AFRL corporate process. However, they should be of a comparable scale and scope as those identified here, be structured to drive advances in key areas identified here as being essential for meeting emerging Air Force needs, and emphasize demonstrations that require sets of individual technology areas to be integrated at the whole-system level.

- Evaluate, define, and focus a set of "grand challenge" problems of sufficient scale and scope to drive major technology development efforts in key areas identified here.
- Structure each "grand challenge" to drive advances in research directions identified in "Technology Horizons" as being essential for meeting Air Force needs in 2030.
- Define specific demonstration goals for each challenge that require sets of individual technology areas to be integrated and demonstrated at the whole-system level.
- Initiate sustained research efforts in AFRL S&T portfolio as necessary to achieve each of the "grand challenge" demonstration goals.

A properly chosen and sustained set of "grand challenges" is a key step in implementing the S&T vision identified in "Technology Horizons", and for advancing a range of technologies that will be essential for meeting Air Force needs.

5. Recommendation #5: Improve Aspects of Air Force S&T Management Process

The final step involves improvements in two key aspects of the processes used by the Air Force for planning and managing its S&T efforts.

5.1. Obtain HAF-Level Endorsement of a Stable S&T Planning Construct

The "Focused Long-Term Challenges" (FLTC) process [see Volume 2] used by AFRL to manage its S&T portfolio was developed to enable effective near-, mid- and long-range planning of research efforts. Some have suggested that a process organized around Service Core Functions should replace this, to better allow coordinating S&T efforts with broader Air Force objectives. Since transition to any new planning construct will entail substantial effort that will not directly contribute to research progress, such change should be done with deliberation and lead to HAF-level endorsement of the resulting planning construct. Air Force S&T will benefit significantly from the stability that such endorsement of an AFRL planning construct can bring, since this can provide the continuity of mid- and long-range planning needed to develop technologies that will be essential a decade or more from now.

5.2. Increase MAJCOM and Product Center Inputs in the AFRL Planning Process

Although the planning construct used by AFRL over the past five years involves regular inputs from all MAJCOMs and Product Centers, the level of involvement in providing such inputs has varied widely. For any AFRL planning construct to be effective in directing S&T to meet Air Force needs, it is essential that processes be established and consistently followed for obtaining such inputs. Regardless of the particular planning construct, this will require a formal process for obtaining periodic inputs from MAJCOMs and Product Centers to make adjustments at higher levels of S&T planning and programming, and a more frequent and less formal process at lower levels. The process for providing such inputs should be designed to ensure an appropriate balance between nearer-term responsiveness to the needs of MAJCOMs and Product Centers and longer-term responsiveness to science and technology opportunities envisioned by AFRL for meeting Air Force needs.

- Obtain HAF-level endorsement of an AFRL planning construct for S&T to provide the stability needed for effective mid- and long-range development of technologies.
- Define and implement a formal process for obtaining high-level inputs from MAJCOMs and Product Centers in periodic adjustments within the AFRL S&T planning construct.
- Develop and implement an informal process to obtain more frequent inputs from MAJCOMs and Product Centers for lower-levels of the AFRL S&T planning construct.

These improvements to the S&T management process will ensure that research efforts over the next decade and beyond will properly account for MAJCOM and Product Center needs to maximize the technology advantage of the U.S. Air Force.

APPENDIX A. TERMS OF REFERENCE

Background

The rapid "flattening" of the world from a technology perspective is allowing science and technology advances made anywhere to be exploited globally for developing militarily significant new capabilities. Many countries already have, or soon will have, the ability to translate worldwide technology advances into new offensive and defensive capabilities in the air, space, and cyber domains, and across domain boundaries. International markets in military systems will diffuse these capabilities rapidly and broadly. As a result, over the next two decades the U.S. will face a growing number of nations having near-peer or peer capabilities, and may find it increasingly difficult to maintain the technology superiority over potential adversaries that it has had in the past. Correctly anticipating those science and technology advances that will have the greatest potential military significance – and the capabilities and counter-capabilities that may be derived from them – can help avoid technology surprise and ensure U.S. capability dominance.

This study will seek to identify key advances in science and technology that are likely to occur over the next 10 years that could in the following 10 years be developed into significant military capabilities. The use of this "10+10 technology-to-capability" forecasting process distinguishes this study from others in the Air Force and elsewhere that aim to understand various aspects of the opportunities and threats that emerging technologies present. Using this process, the study will develop a forward-looking yet realistic assessment on a 20-year horizon of potential offensive and defensive capabilities and counter-capabilities of the Air Force and its possible future adversaries.

Study Products

Briefing to SAF/OS & AF/CC in December 2009. Publish report in February 2010.

Charter

The study will:

- Conduct a "next-decade" (2020) assessment of technology advances that will be key to future air, space, and cyber domain capabilities, and to potential cross-domain capabilities.
- Provide a "following-decade" (2030) assessment of U.S. and adversary capabilities that could be developed from these technology advances, focusing on potential

"leapfrog" and "game-changing" capabilities that may substantially alter future warfighting environments.

- Determine counter-capabilities that the Air Force will need in 2030 to be effective against these potential new adversary capabilities.
- Identify the underlying technologies that the Air Force will need in 2020 in order to develop the counter-capabilities it needs in 2030.
- Identify the science and technology research efforts that the Air Force must start today to develop the technologies it needs in 2020 to obtain the counter-capabilities it needs in 2030.

APPENDIX B – "TECHNOLOGY HORIZONS" SOURCES OF INPUTS

This Appendix describes the principal sources of inputs that were considered in developing the findings and recommendations in "Technology Horizons".

B.1. "Technology Horizons" Working Groups

One source of inputs to "Technology Horizons" was a set of working groups specifically formed to provide a broad range of information, ideas, and viewpoints to this effort.

B.1.1. Role of the Working Groups

Informal working groups were assembled for each of the air, space, and cyber domains, and an additional working group was assembled to address cross-domain topics. Each group served as a forum in which technical opinions of experts were provided to AF/ST and were discussed for possible further consideration in later phases of the effort, along with data and inputs from other sources. All working group discussions were conducted at Secret level. Additional inputs at higher levels of classification were obtained by other means in later phases of the effort.

Based on guidance from SAF/GC, to permit the free exchange of ideas and viewpoints that was essential to developing "Technology Horizons" while remaining consistent with federal rules that would otherwise apply to formal advisory committees, all working groups were formed on an *ad hoc* basis and were managed in a way that allowed sharing of individual opinions while specifically avoiding all efforts to reach any formal consensus among the group participants. Participants were involved in a strictly individual capacity; the opinions they expressed were their own views and did not necessarily represent the views of their organization. In addition, participants were specifically instructed not to discuss any contractor proprietary information. Discussions were of a broad cross-cutting nature, and no current contracts or specific Air Force requirements were discussed. There was no access to source selection information, contractor bid information, or proposal information. Operating the working groups in this manner gave the benefits of a broad range of views among the inputs to "Technology Horizons" without the rigid requirement of formal consensus that would have applied to formal advisory committees, which would have risked limiting the insights from these groups to largely predictable views.

B.1.2. Composition of the Working Groups

Participants for the working groups were selected in part for the combination of their scientific and technical background and knowledge, their direct experience and knowledge of Air Force S&T and Air Force operations, and their ability to contribute effectively in a fast-paced idea intensive group environment. Participants were also selected to achieve reasonable coverage of the technical areas relevant to each domain, while limiting the size of each group to roughly a dozen participants to allow for exchange of ideas and opinions. Working group participants were selected from the:

- Air Force S&T community
- Intelligence community
- MAJCOMs
- Product Centers
- FFRDCs
- Defense industry
- Academia

Since all participants strictly represented their own opinions and not their organization's views, no attempt was made to achieve any specific target balance among types of organizations or to seek comprehensive representation of all organizations within any of these sectors.

B.1.2.1. Air Domain Working Group

This Working Group considered technologies, concepts, and systems relevant to air domain functions, such as remotely piloted aircraft, fractionated strike system architectures, electronic warfare, bandwidth, GPS-denied environments, air domain threat systems, LO and CLO technologies, sensors, ISR systems, low collateral damage munitions, air mobility, lighter-than-air and hybrid airships, base defense and force projection, power and thermal management, airbreathing hypersonic systems, near-space systems, and numerous other topics relevant to the air domain.

- Dr. Werner J.A. Dahm, Chair
- Col Eric Silkowski, Ph.D., Vice Chair
- Dr. Edward H. "Ned" Allen
- Douglas L. Bowers
- Dr. Gregory K. Crawford
- Dr. Donald R. Erbschloe
- Dr. Janet S. Fender
- Col Robert S. Fredell, Ph.D.
- Dr. Mark A. Gallagher
- Dr. Thomas T. Hamilton
- Dr. Brian M. Kent
- Dr. Michael A. Kuliasha
- Betsy S. Witt
- Dr. John C. Zolper

B.1.2.2. Space Domain Working Group

This Working Group examined technologies and systems for space launch, communications, space system architectures, precision navigation and timing, space situational awareness, space protection, orbit transfer and maneuver, space-based intelligence, surveillance and reconnaissance, space radar, missile warning and intercept, space weather, space debris, small satellites, responsive space, laser communications, and other topics relevant to space domain functions.

- Dr. Werner J.A. Dahm, Chair
- Col. Eric Silkowski, Ph.D., Vice Chair
- Dr. William L. Baker
- Dr. Alok Das
- Dr. Roberta Ewart
- Dr. Jonathan Gordon
- Dr. David A. Hardy
- Dr. Daniel Hastings
- Dr. Michael A. Kuliasha
- Gary A. O'Connell
- Dr. Jim F. Riker
- Dr. Lara S. Schmidt
- Dr. Dwight C. Streit

B.1.2.3. Cyber Domain Working Group

This Working Group examined offensive and defensive technologies and concepts relevant to the cyber domain, such as cyber situational awareness, cyber forensics and attribution, data protection, polymorphic networks, mobile ad hoc networks, information assurance, SCADA systems, honeypots, deterrence, continuity of operations, mission assurance, cyber agility, attack containment, recovery, fight-through approaches, encryption, quantum key distribution, quantum computing, and other considerations relevant to the cyber domain.

- Dr. Werner J.A. Dahm, Chair
- Col. Eric Silkowski, Ph.D., Vice Chair
- Dr. John S. Bay
- Richard J. Byrne
- Dr. Chris Colliver
- Jon Goding
- Dewey Houck
- Jeff A. Hughes
- Dr. Kamal T. Jabbour
- Dr. Michael A. Kuliasha
- Richard Mesic
- Dr. Marc Zissman

B.1.2.4. Cross-Domain Working Group

This Working Group considered technologies and systems that (i) have expected effects spanning across two or more of the air, space, and cyber domains, (ii) have unexpected effects in a domain different other that in which expected effects occur, (iii) may require key supporting functions from another domain, or (iv) fall between these classical domains but have implications in one or more of these.

- Dr. Werner J.A. Dahm, Chair
- Col. Eric Silkowski, Ph.D., Vice Chair
- Dr. Edward H. "Ned" Allen
- Dr. William L. Baker
- Richard J. Byrne
- Dr. Alok Das
- Dr. Richard P. Hallion
- Jeff A. Hughes
- Dr. Brian M. Kent
- Dr. Michael A. Kuliasha
- Dr. Jim F. Riker

B.1.3 Working Group Participant Biographies

The biographical summaries of the Working Group participants below give an overview of their respective expertise and background.

Dr. Edward H. "Ned" Allen

Dr. Allen is Chief Scientist of the Skunk Works at Lockheed Martin Aeronautics Company's Advanced Development Programs in Palmdale, CA, where he is also a Corporate Senior Fellow. In his role as Chief Scientist he represents the Skunk Works in various forums and serves as senior advisor for science and technology across the Corporation. He is the principal inventor on more than two dozen patents, and currently conducts research in areas that include energy, quantum information, hypersonic systems, hybrid airships, and micro air vehicles. After a decade on the tenured faculty of Utah State University and two years as a Rockefeller Foundation Fellow, he founded and ran his own commuter airline, now called Horizon Air, and then founded Daedalus Research Inc., an aeronautical research and development firm. In 1998, he brought his firm's key assets into Lockheed Martin's Skunk Works. He has degrees in physics from Stanford University, in mathematics and literature from Swarthmore College, and in applied mathematics and international relations from the University of Pennsylvania.

Dr. William L. Baker

Dr. Baker was Chief Scientist of the Directed Energy Directorate in the Air Force Research Laboratory (AFRL), where as an internationally recognized leader in the entire spectrum of directed energy technologies he has led research and development on high-energy laser weapons technology and the application of advanced optics to space situational awareness. He served as an Air Force nuclear research officer prior to becoming a civilian scientist at the Air Force Weapons Laboratory. He initially performed and led research on high-energy plasmas and pulse power systems for simulating nuclear weapon effects, and then began work on high-energy particle beams for directed energy weapon technologies. He led a joint effort to develop a unique accelerator that demonstrated stable beam propagation in open air. He then created and led the Air Force high power microwave weapon technology program, which became the center of excellence for the nation and has demonstrated multiple military applications. He is a Fellow of IEEE and president of the Directed Energy Professional Society. He has received the DOD Distinguished Civilian Service Award and the Distinguished Presidential Rank Award. He has B.S. and M.S degrees in physics from Ohio State University, and a Ph.D. in nuclear physics from Ohio State University.

Dr. John S. Bay

Dr. Bay served as Chief Scientist of the Information Directorate in the Air Force Research Laboratory (AFRL), directing the planning and technical execution of information systems science and technology and its transition to air, space, and cyberspace systems. Prior to joining AFRL he was a Program Manager in the Information Exploitation Office (IXO) at DARPA, an Engineering Fellow at Raytheon Company, and a professor of electrical and computer engineering at Virginia Tech. His research focus is on control systems, robotics, biomechanics, machine learning, and embedded systems. He is an author on more than 65 publications, a Fellow of the Institute of Electrical and Electronics Engineers (IEEE), and is a winner of the 2009 IEEE Computer Society Technical Achievement Award for his work in embedded systems. He is a former IEEE Computer Society Distinguished Visitor and is a 2005 recipient of the Office of the Secretary of Defense Medal for Exceptional Public Service. He holds a B.S. degree in electrical engineering from Virginia Tech and M.S. and Ph.D. degrees from Ohio State University.

Douglas L. Bowers

Mr. Bowers is Director of the Propulsion Directorate in the Air Force Research Laboratory, where he directs the Air Force's science and technology program in propulsion and power for space, missile and aircraft applications. He previously was Associate Director for Air Platforms in AFRL's Air Vehicles Directorate, responsible for fixed wing and rotary wing technologies, hypersonics, turbine engines, and power. He has served in a variety of senior technical positions within the Air Force, leading the development and fielding of advanced highly survivable engine inlets and exhaust nozzles. He was a technical consultant on the C-17, F-15E, F-16 and B-1 development programs. He has served as a member and U.S. national coordinator for the NATO Advisory Group for Aerospace Research and Development (AGARD) fluid dynamics panel. As the sole Air Force staff member for the Commission on the Future of the US Aerospace Industry, he developed recommendations for aerospace industry health and national security. He has a B.S. degree in aerospace engineering from Purdue University, an M.S. degree in aerospace engineering from Ohio State University, and an M.S. degree in engineering management from the University of Dayton.

Richard J. Byrne

Mr. Byrne is a Vice President at The MITRE Corporation with nearly 30 years of experience developing computing, communications, networks, information technologies, and systems engineering products and prototypes. His responsibilities include technical centers, research programs, and other engineering activities in the Command and Control Center (C2C), exploring solutions to national security problems with emphasis on improved information interoperability, systems integration, and cyber assurance, including net-centric strategies, complex systems engineering, and information technologies. He has held various positions at MITRE supporting the Air Force and DoD, including vice-president for all Air Force programs and previously technical director for the Air Force Electronic Systems Center and executive director of innovation. He was a founder and engineering manager at a semiconductor startup, leading design and production release of more than 150 products over five years. He was technical manager of VLSI design methods for telecommunication at ITT's Advanced Technology Center. He has a B.S. in electrical engineering and M.S. in electrical engineering and computer science from MIT.

Dr. Chris Colliver

Dr. Colliver is the Technical Advisor for C4/IO Analysis at the National Air and Space Intelligence Center, where he leads intelligence analysis and production of foreign C4ISR and Cyberspace capabilities to support joint operational, acquisition and policy making. He has worked in a number of analytic, advisory, and supervisory positions responsible for foreign space and counterspace systems analysis. He established a Counterspace Analysis Division and expanded NASIC's Defense Intelligence Space Order of Battle (DISOB) to help lead the transformation of NASIC into a more space-centric organization. He served as the Space Topic Manager for the National Intelligence Priorities Framework from 2004-2008, and was named an Exceptional Intelligence Analyst by the Director of Central Intelligence. He has experience in analog and digital communications via HF, microwave, and satellite, as well as network operations. He holds a B.S. degree, an M.S. degree, and a Ph.D degree, all in Electrical Engineering from the University of Dayton.

Dr. Gregory K. Crawford

Dr. Crawford is the Director for Integration of ISR at MITRE, focused on ISR sensing, sensor data and signal processing, exploitation algorithm development and prototyping, and enterprise systems engineering. He also leads work on integrated sensing, processing and exploitation as part of the MITRE Corporate Challenge, in which he guides 30 research and development projects focused on operationally motivated capabilities directed at difficult problems within this domain. He has led multiple projects with contributions ranging from basic R&D to system design and end-to-end systems engineering. Recently he served as Project Leader and Chief Engineer for Space Radar (SR) Battle Management Command and Control (BMC2), addressing a wide variety of ISR exploitation and enterprise-level systems engineering issues seeking to develop "dual-use" (Tactical-National) capabilities. He is an author of more than 20 papers on a broad array of Geophysical topics in leading journals and has written numerous technical papers on applied topics. He has a B.S. in Physics and B.A in Chemistry from California State University, Fresno, an M.S. in Geophysics and Space Physics from UCLA, and a Ph.D. in Space Plasma Physics from UCLA.

Dr. Werner J.A. Dahm

Dr. Dahm is the Chief Scientist of the U.S. Air Force. He is the principal advisor for science and technology to the Air Force Chief of Staff and the Secretary of the Air Force. He is on leave from the University of Michigan, where for the past 25 years he has served as a Professor of Aerospace Engineering. His work has principally been in areas related to aerodynamics and propulsion. He is an author of over 180 journal articles, conference papers, and technical publications, a holder of several patents, and has given over 120 invited, plenary, and keynote lectures worldwide on topics dealing with aerospace engineering. He has been a member of the Air Force Scientific Advisory Board (SAB) and has served on numerous task forces for the Defense Science Board (DSB) and as a member of the Defense Science Study Group (DSSG). He is a Fellow of both the American Physical Society (APS) and the American Institute of Aeronautics and Astronautics (AIAA), and a recipient of the William F. Ballhaus Aeronautics Prize from Caltech as well as major research awards from the University of Michigan. He has served extensively in advisory and organizational roles in his field, and as a consultant for industry. He has B.S. and M.S. degrees in mechanical engineering from the University of Alabama and University of Tennessee Space Institute, and a Ph.D. degree in aeronautics from Caltech.

Dr. Alok Das

Dr. Das is Senior Scientist for Design Innovation in the Air Force Research Laboratory. As chief innovation officer at AFRL, he is the principal advisor for technology and process innovation strategies, and leads the rapid reaction team that utilizes innovation and collaboration to develop near-term solutions to urgent warfighter needs. He previously was Chief Scientist in AFRL's Space Vehicles Directorate, focusing on revolutionary space mission architectures, satellite designs, and technologies to reduce cost and increase capability and operating flexibility of space systems. He first was an engineer in the Indian Space Research Organization, then joined the Air Force Rocket Propulsion Laboratory to become technical lead for large space structures. He was instrumental in developing and transitioning smart structures technology from conceptual demonstrations to applications in space systems, giving satellite designers an array of space-proven technologies. He is a Fellow of the American Institute of Aeronautics & Astronautics and an author of more than 60 technical articles on space technologies. He has a B.S. degree in electronics and communications engineering and an M.S. in aeronautical engineering from the Indian Institute of Science in Bangalore, and a Ph.D. in aerospace engineering from Virginia Tech.

Dr. Donald R. Erbschloe

Dr. Erbschloe is the Chief Scientist for Air Mobility Command (AMC), where he advises the AMC Commander and key leadership on scientific issues and technological enhancements of AMC capabilities. He has had a 28-year military career in the Air Force balanced among three primary thrusts: operations, academia, and scientific and technical management. He was a command pilot with 3,900 flying hours in the C-141, the TG-7A, and the UV-18. He has served as an Associate Professor of physics at the Air Force Academy, as Chief Scientist of the European Office of Aerospace Research and Development (EOARD) in AFOSR, and as Director of Faculty Research at the Air Force Academy. He later served as Military Assistant to the Chief Scientist of the Air Force, and most recently was acting Chief Operations Officer for the Office of Science in the Department of Energy, which manages 10 world-class laboratories and is the single largest supporter of basic research in the physical sciences in the United States. He has a B.A. degree in physics and mathematics from the University of Virginia, an M.S. degree in physics from the Air Force Institute of Technology, and a Ph.D. degree in physical electronics from Oxford University in England.

Dr. Roberta M. Ewart

Dr. Ewart is the Chief Scientist for the Air Force Space and Missile Systems Center in El Segundo, CA. Her responsibilities include the identification, maturation and transition of key technologies for National Security Space missions. Her career experience spans space flight operations with NASA and the USAF, as well as research, development and acquisitions positions. Her research focus areas include solid-state lasers, photonics, laser communication, and dynamic radiation response modeling of spacecraft. She was named the 2008 Air Force Space Command Senior Civilian Manager of the Year. She holds a B.S. degree in physics from the United States Air Force Academy, degrees in Theoretical Physics and Philosophy from Oxford University, and a degree in Electrical Engineering from Stanford University.

Dr. Janet S. Fender

Dr. Fender is the Chief Scientist in Air Combat Command (ACC), serving as scientific adviser to the Commander and providing scientific expertise and technical guidance throughout ACC. She serves as the primary interface for ACC to the scientific community to identify technologies and catalyze enhancements to ACC's warfighting capabilities. Her career in science and technology includes basic laboratory research, fielding new capabilities, key technical positions in major programs, and support to operations. Recently she led Air Force studies on "Day Without Space" and "The Value of Stealth". She previously was Chief Scientist for the Space Vehicles Directorate in AFRL and Senior Scientist for Directed Energy and Advanced Imaging. She has authored over 100 technical papers, is a Fellow of the Optical Society of America and the International Society of Optical Engineers, and serves on the Executive Committee of the Board of Directors of the American Institute of Physics. She has a B.S. degree in physics and astronomy from University of Oklahoma, and M.S. and Ph.D. degrees in optical sciences from the University of Arizona.

Col Robert S. Fredell, Ph.D.

Colonel Fredell is Chief Scientist and Director of Research at the U.S. Air Force Academy, where he directs thirteen scientific and engineering research centers and two institutes in the largest undergraduate university research program in the U.S. He has performed basic and applied research in aircraft structural fatigue and composite structural repair, and has taught engineering mechanics at the Academy. He has expertise in advanced metallic and composite materials, with particular focus on fiber metal laminates and their application to new and aging aircraft. He previously served as the Military Assistant to the Air Force Chief Scientist in the Pentagon, as Deputy Director of Engineering at Warner Robbins ALC, as Director of Faculty Research at USAFA, and as Technical Director of the European Office of Aerospace Research and Development (EOARD) of the Air Force Office of Scientific Research (AFOSR). He has B.S. and M.S. degrees in mechanical engineering from Oklahoma State University, and a Ph.D. degree in aerospace engineering from Delft University in the Netherlands.

Dr. Mark A. Gallagher

Dr. Gallagher is Technical Director in Studies & Analyses, Assessments, and Lessons Learned (AF/A9) in Headquarters Air Force. In this role he supports analyses across the spectrum of Air Force missions, and develops and applies modeling and simulation to support Air Force warfighting and force structure capability and sufficiency assessments. He previously was a member of the full-time faculty at the Air Force Institute of Technology (AFIT), where he continues to serve as an adjunct Associate Professor. He has had prior assignments in the Cost Analysis Improvement Group (CAIG) of Program, Analysis and Evaluation (PA&E) in the Office of Secretary of Defense (OSD), and as Chief of Missile Defense and Combating WMD Capabilities Division at U.S. Strategic Command. He founded the Information Operations Joint Munitions Effectiveness Manual, which he also chaired for four years. He is serving a second term on the Military Operations Research Society (MORS) Board of Directors as Vice-President of Society Services. He holds B.S. degrees in operations research and computer science from the U.S. Air Force Academy, and M.S. and Ph.D. degrees in operations research from the Air Force Institute of Technology (AFIT).

Jon Goding

Mr. Goding is a Principal Engineering Fellow with Raytheon's Network Centric Systems business. He presently serves as chief engineer for Raytheon's Cyber Initiative, where he is responsible for coordinating research and development in computer network operations technology across the company. In 23 years at Raytheon, he has designed computing and network systems for many DoD and Federal Government customers, and was the architect for information assurance on the Navy Marine Corps Internet from pre-award through initial operations. At the time it went operational, this was the largest integrated secure network in use. When Raytheon formed a new Secure Networks product line in 2003, he was named its Technology Director. During recent years he has participated in many government and industry working groups, and sponsored and advised numerous university research projects. He has a B.S. degree in electrical engineering from the University of Florida, and an M.S. degree in electrical engineering from the University of South Florida.

Jonathan D. Gordon

Mr. Gordon is a Program Area Chief Engineer for the Advanced Concepts and Technology group at Raytheon Space and Airborne Systems (SAS), with engineering oversight for all the company's RF technology development programs and internal research and development efforts in this business area. He is also responsible for the development of long range and "white space" technology roadmaps that guide internal investment and program pursuits. He has 30 years of technical and programmatic leadership at Raytheon and Hughes Aircraft Company. His primary expertise is active array subsystem design, integration and test for both space and airborne applications. Prior to his Chief Engineer assignment, he was the manager of Advanced RF Space Programs in SAS, where he was involved in numerous space RF programs including the ESA/FEP Technology Development Program (Space Based Radar), Innovative Space Antenna Technology (ISAT), TechSAT 21, and Discoverer II, and has served as a technical and IPT lead for several active array subsystem engineering efforts. He has B.S. degrees in physics and in electrical engineering from University of Southern California (USC), and an M.S. degree in electrical engineering from USC.

Dr. Richard P. Hallion

Dr. Hallion is an internationally recognized aerospace historian and founder of Hallion Associates, which evaluates emerging technologies and assesses their potential for application to meet civil and military aerospace needs. He retired from the United States Air Force in 2006 as Senior Advisor for Air and Space Issues, served also as the Air Force Historian for over a decade, and held the Harold Keith Johnson Visiting Chair in Military History at the U.S. Army Military History Institute, Army War College. He was a founding Curator of the National Air and Space Museum, and subsequently held the Charles A. Lindbergh Chair in Aerospace History, and an Alfred Verville Fellowship in Aeronautics. He has written over a dozen works on the history of flight and military and civil aviation, teaches and lectures widely, and is a Fellow of the Earth Shine Institute and a Fellow of the American Institute of Aeronautics and Astronautics (AIAA). He holds a B.A. degree and a Ph.D. degree, both from the University of Maryland.

Dr. Thomas T. Hamilton

Dr. Hamilton is Senior Physical Scientist in national security at the RAND Corporation, where since 1996 he has worked on a variety of issues for the Department of Defense, with a focus on unmanned aircraft, airport security, aircraft test facilities, and force modernization. In particular, his work has addressed the future of USAF technology, aircraft survivability issues, UAS procurement and operations, USAF tanker modernization, force modernization and employment, aircraft test facilities, and airport security. Prior to joining RAND, from 1986 to 1996 he was a research astrophysicist at Harvard University, Columbia University, and the California Institute of Technology, and from 1981 to 1983 he developed satellite guidance software at Martin Marietta Corporation. He also served on the faculty in the Department of Astronomy at Caltech. He has an A.B. degree in physics from Harvard University and a Ph.D. in physics from Columbia University.

Dr. David A. Hardy

Dr. Hardy is Associate Director for Space Technology in the Space Vehicles Directorate of the Air Force Research Laboratory (AFRL). He is responsible for oversight of space science and technology investments and for the senior civilian management of the Space Vehicles Directorate's science and technology portfolio. He leads AFRL's efforts across the full range of existing and emerging space technologies needed to support present and future space systems for the Air Force and Department of Defense, including science and technology related to sensors, materials, information technology, space vehicles, directed energy, propulsion and human effectiveness. He also serves as an adviser on space science and technology to Air Force Space Command, the Space and Missile Systems Center, DARPA, and other organizations, and develops and maintains coordination in the growth and execution of space science and technology by national and international governmental agencies and academia. He has a B.S. degree in physics from Duke University, and M.S. and Ph.D. degrees in space physics and astronomy from Rice University.

Dr. Daniel Hastings

Dr. Hastings is a Professor of Aeronautics and Astronautics at the Massachusetts Institute of Technology, and serves as MIT's Dean for Undergraduate Education. He has taught courses in plasma physics, rocket propulsion, space power and propulsion, aerospace policy, and space systems engineering. He was the Chief Scientist of the U.S. Air Force from 1997 to 1999, serving as the chief scientific adviser to the Air Force Chief of Staff and Secretary of the Air Force. He led several influential studies on where the Air Force should invest in space, global energy projection, and options for a science and technology workforce for the 21st century. His research is on space systems and space policy, and has also focused on issues related to spacecraft environmental interactions, space propulsion, space systems engineering, and space policy. He is a Fellow of AIAA, a Fellow of INCOSE, a member of the International Academy of Astronautics, a past member of the National Science Board, and former chair of Air Force Scientific Advisory Board. He has a B.A. in mathematics from Oxford University in England, S.M. and Ph.D. in aeronautics and astronautics from MIT.

Dewey R. Houck, II

Mr. Houck is a Senior Technical Fellow working in Boeing's Integrated Defense Systems for the Mission Systems organization, and since July 2008 has been the Director of the Mission Systems organization. He previously served as the Chief Technical Officer for Mission Systems with primary responsibilities in the areas of technology strategy development, research planning and integration of Mission Systems Intellectual Property into other Boeing programs. Mission Systems is the Boeing business unit that serves intelligence community customers including the NSA, CIA, NGA and DNI, with its focus on developing and deploying information management systems to facilitate tasking, collection, sharing and analysis of intelligence. He previously served on the senior management team of Autometric, Inc. as Vice President for Technology Development, where he administered all research and product development activities including several geospatial, photogrammetric and visualization initiatives. He has B.S. and M.S. degrees in civil engineering with specialization in photogrammetry and geodesy from Virginia Tech.

Jeff A. Hughes

Mr. Hughes is Chief of the Autonomic Trusted Sensing for Persistent Intelligence (ATSPI) Technology Office in the Sensors Directorate of the Air Force Research Laboratory (AFRL), where he leads research on complex system decomposition, vulnerability, and risk analysis to develop trusted systems. He was selected to establish the ATSPI Technology Office to support the OSD Software Protection Initiative (SPI) to prevent unauthorized distribution and exploitation of application software and associated data critical to national security. ATSPI is the field office for the DoD Anti-Tamper Executive Agent with the mission to deter the reverse engineering and exploitation of the U.S. military's critical technologies. His prior work has included developing and deploying combat identification capabilities, and low observable/counter-low observable radio frequency and electronic warfare technology to Air Force weapons systems. He has served on numerous NATO, US Government inter-agency committees, and DoD working groups, and has received numerous awards for his work. He has B.S. and M.S. degrees in electrical engineering from Ohio State University, and has completed work toward his Ph.D. at the Air Force Institute of Technology.

Dr. Kamal T. Jabbour

Dr. Jabbour is Senior Scientist for Information Assurance in the Information Directorate of the Air Force Research Laboratory (AFRL), where he serves as principal scientific advisor for information assurance, including defensive and offensive information warfare and technology. He plans and advocates major research and development activities, guides the quality of scientific and technical resources, and provides expert technical consultation to organizations across the Air Force, DoD, government agencies, universities and industry. He is Program Director of the Advanced Course in Engineering Cyber Security Boot Camp for developing future Air Force cyber warfare officers and leaders. He began his professional career on the computer engineering faculty at Syracuse University, where he taught and conducted research for two decades, including a three-year term as department chairman. In 1999, he joined the Cyber Operations Branch at AFRL through the Intergovernmental Personnel Act, and transitioned gradually from academia to his current position. He has a B.S. degree in electrical engineering from American University of Beirut in Lebanon, and a Ph.D. degree in electrical engineering from University of Salford in England.

Dr. Brian M. Kent

Dr. Kent is Chief Scientist of the Sensors Directorate in the Air Force Research Laboratory (AFRL), where he serves as principal scientific and technical adviser with primary authority for the technical content of the sensors research portfolio, identifying gaps and analyzing advancements in a variety of scientific fields. From 1985 to 1992 he conducted classified research that made pioneering contributions to signature measurement technology and established international standards for radar signature testing. He previously worked in the Passive Observables Branch of the Avionics Laboratory, and later in the AFWAL Signature Technology Office. He is an author of more than 85 archival articles and technical reports and has written key sections of classified textbooks and design manuals. He has delivered more than 200 lectures, and developed a DoD Low Observables Short Course that has been taught to more than 2,000 scientists and engineers since 1989. He has served as a technical consultant to a wide range of federal agencies, as an international technical adviser for the DoD, and has provided basic research guidance to leading academic institutions. He has a B.S. degree in electrical engineering from Michigan State University, and M.S. and Ph.D. degrees in electrical engineering from Ohio State University.

Dr. Michael A. Kuliasha

Dr. Kuliasha is the Chief Technologist at the Air Force Research Laboratory (AFRL), where he is the principal technical adviser to the AFRL Commander on the Air Force's $2 billion science and technology program and the additional $2 billion customer-funded research and development conducted at AFRL. As Chief Technologies he also provides technical guidance for the AFRL workforce of approximately 10,000 people. Prior to his current role he has had more than 30 years of experience in the Department of Energy's Oak Ridge National Laboratory, where he has held a variety of leadership positions including Director of the Computational Physics and Engineering Division, Director for Homeland Security, and acting Associate Laboratory Director for Computing, Robotics and Education, among others. He has technical expertise in a broad-range of scientific disciplines including nuclear science and technology, high-performance computing, modeling and simulation, knowledge discovery, and energy technologies. He holds a B.S. degree in mathematics and M.S. and PhD. Degrees in nuclear engineering from the University of New Mexico.

Richard Mesic

Mr. Mesic is a Senior Policy Analyst at RAND Corporation with over 39 years of professional experience in requirements definition, system and operational concept development, and quantitative evaluation. His recent work has focused on cyber security, irregular warfare, counter-terrorism, and C4ISR/space system studies, including information operations and assurance, national intelligence systems, critical infrastructure protection, counter IED, and special operations. He has studied Air Force operations in all recent conflicts, and was Principle Investigator on the Project Air Force study in support of the Air Force AFCYBER initiative. He continues to co-lead a multi-year study for the Navy on advanced C4ISR systems for littoral counter-terrorism operations. His prior work includes strategic nuclear deterrence and systems, concepts to counter weapons of mass destruction, ballistic missile defense, arms control and arms control verification requirements and capabilities, strategic and tactical C4ISR, national-level intelligence issues, space, information operations, and counter-proliferation policy. He has a B.A. degree in mathematics from Knox College and an M.S. degree in mathematics from Michigan State University.

Gary A. O'Connell

Mr. O'Connell is Chief Scientist in the National Air and Space Intelligence Center (NASIC), the Air Force and DoD Center of Excellence for all-source air and space intelligence. He has over 30 years of experience in aerospace and intelligence fields ranging from electronic countermeasures and tactics development to drafting National Intelligence Estimates for national policymakers. As Chief Scientist at NASIC he is senior advisor to the Commander and oversees analytic efforts to support Air Force and Joint operational, acquisition, and policymaking customers in the national intelligence community, and serves on numerous panels and committees in the national intelligence and scientific communities. He began his career as an air-to-air missile analyst in the Foreign Technology Division, and has held various management positions until being appointed Chief Scientist for Global Threat in 1997. He has served as the Air Force representative on the Weapon and Space Systems Intelligence Committee's Aerodynamic Missile Systems Subcommittee, and as the Air Force representative on the Scientific and Technical Intelligence Committee from 2001 to 2004. He has a B.S. degree in aerospace engineering from the University of Cincinnati and an M.S. degree in national resource strategy from National Defense University.

Dr. Jim F. Riker

Dr. Riker is the Chief Scientist of the Space Vehicles Directorate in the Air Force Research Laboratory (AFRL), where he directs technical research in space situational awareness (SSA), defensive and offensive space control, responsive space, intelligence, surveillance, and reconnaissance (ISR), satellite communication and satellite control. He has also served in a variety of research roles in optical surveillance and space-based laser system program offices. He is internationally recognized for contributions in the areas of laser beam control, atmospheric propagation, atmospheric compensation, active and passive tracking, and imaging, and is an author of more than 50 scientific and professional publications. He previously was Technical Director for the Optics Division of the Directed Energy Directorate in AFRL, and led the Active Track program that produced numerous first-ever technology demonstrations for the MDA missile defense system. He is a Fellow of the Society for Photo-optical and Instrumentation Engineering (SPIE), and the 2007 winner of the Air Force Harold M. Brown Award. He has B.S. in physics and mathematics from New Mexico Institute of Technology, and an M.S. and Ph.D. in physics from the University of New Mexico.

Dr. Lara S. Schmidt

Dr. Schmidt is a Senior Statistician and Defense Analyst at RAND, where her research focus is military space systems, GPS, precise time and frequency systems, atomic timekeeping, and risk assessment. Her recent work includes assessments of DOD space systems in hostile environments, space control capabilities, space vulnerability and dependence, weapon system use of space systems, operational level C2, military PNT systems, Air Force special operations, irregular warfare, counterspace threats, and risk to warfighters from attacks on space systems. She is RAND liaison to Air Force Space Command, is on the advisory board for the DoD precise timing organization, serves as a referee for several technical journals, and has held leadership positions in the American Statistical Association, including chair of the section on Defense and National Security. Prior to joining RAND she spent eight years as a government civilian working with GPS and atomic timekeeping. She holds a B.S. in mathematics from Shepherd College, an M.S. in mathematics from West Virginia University, and a Ph.D. in mathematics from American University.

Col Eric Silkowski, Ph.D.

Col Silkowski is the Military Assistant to the Chief Scientist of the U.S. Air Force in the Pentagon. He supports the Chief Scientist in providing independent, objective and timely scientific and technical advice to the Air Force Chief of Staff and the Secretary of the Air Force, and in evaluating technical issues of relevance to the Air Force mission. He also supports the Chief Scientist in his contributions to supporting and maintaining the technical quality of the research being conducted across the Air Force. Previously he led the Air Force Technical Applications Center's (AFTAC) Applied Physics Laboratory and ran worldwide operations for nuclear event detection and global atmospheric monitoring. Other technical assignments include high explosive testing, ballistic re-entry vehicle acquisition, and conceptual design of directed energy weapons systems. He has also served as executive officer to the J8 for NATO Allied Command Operations at Supreme Headquarters Allied Powers Europe. He has a B.A. degree in physics from the University of Chicago, and has M.S. and Ph.D. degrees in engineering physics from the Air Force Institute of Technology.

Dr. Dwight C. Streit

Dr. Streit is the Vice President for Electronics and Sensors at Northrop Grumman Aerospace Systems. In this role he is responsible for all technology development required for advanced semiconductors, microelectronics, radars, communication systems and satellite payload electronics, and sensors and electronic systems for airborne and space-based platforms. Prior to joining Northrop Grumman via its acquisition of TRW, he was president of the TRW company that manufactured high-performance chips for fiber-optic and wireless communication systems. He has more than 300 technical publications and conference presentations, and some 30 patents issued or pending. He was made a member of the National Academy of Engineering for his contributions to the development and production of heterojunction transistors and circuits. He is a Fellow of the Institute of Electrical and

Electronics Engineers (IEEE), a member of the NASA Space Foundation Technology Hall of Fame, and a Fellow of the American Association for the Advancement of Science. In 2003 he was named the UCLA Engineering Alumnus of the Year. He has B.S. degrees in engineering and chemistry from California State University, and M.S. and Ph.D. degrees in electrical engineering from the University of California, Los Angeles (UCLA).

Betsy S. Witt

Ms. Witt is Technical Director of the Air and Cyberspace Analysis Group of the National Air & Space Intelligence Center (NASIC), the Air Force and Department of Defense Center of Excellence for all-source air and space intelligence to support USAF and Joint operational, acquisition, and policymaking customers in the national intelligence community. The Air and Cyberspace Analysis Group is the intelligence community focal point for analyses of foreign air and air defense systems, and for analyses of cyber intrusions into Air Force computer networks. In her capacity as Technical Director she provides technical oversight of analysis on foreign fixed-wing fighters and air armaments, UAVs, aircraft avionics, ground-based early warning radars, command and control, integrated air defense systems, and foreign computer network attack. She began her career in the Foreign Technology Division and has over 31 years experience in intelligence community assignments with the Air Force. She has B.S. and M.S. degrees in mathematics from Wright State University, as well as extensive undergraduate and graduate level courses in engineering.

Dr. Marc A. Zissman

Dr. Zissman is Assistant Head of the Communications and Information Technology Division at MIT Lincoln Laboratory. He leads the Laboratory's cyber protection activities and shares responsibility for research, development, evaluation, and technology transfer in the areas of advanced military communications, wideband tactical networking on-the-move, and language processing. He joined the Laboratory in 1983, with his early research focused on digital speech processing, including parallel computing for speech coding and recognition, co-channel talker interference suppression, language and dialect identification, and cochlear-implant processing. After his subsequent research interests expanded to include information assurance technologies, he served for four years as a U.S. technical specialist to the NATO task group that studies military applications of speech technology, and four years on the Speech Processing Technical Committee of the IEEE Signal Processing Society. He also served four years on the Defense Advanced Research Projects Agency (DARPA) Information Science and Technology Study Group. He holds S.B., S.M., and Ph.D. degrees in electrical engineering from the Massachusetts Institute of Technology.

Dr. John C. Zolper

Dr. Zolper is Vice President, Corporate Research & Development, and Deputy for Corporate Technology and Research for Raytheon Company. In this role he is responsible for strategic technology planning and technology innovation across the enterprise, through which he has been involved in development and execution of an integrated enterprise-wide technology and research vision and strategy. Before joining Raytheon he was the Director of DARPA's Microsystems Technology Office, where he was responsible for the strategic planning and execution of research programs covering all areas of advanced component technology, including electronics, photonics, MEMS, algorithms, and component architecture. Prior to joining DARPA he was a program officer at the Office of Naval Research, where he managed a portfolio of basic and applied research in advanced electronics, was a Senior Member of Technical Staff at Sandia National Laboratories, and was a post-doctoral fellow conducting research on high efficiency silicon solar cells at University of New South Wales in Australia. He is a Fellow of the IEEE and AIAA. He has a B.A. in physics from Gettysburg College, and MEE and Ph.D. degrees in electrical engineering from the University of Delaware.

B.2 Site Visits, Briefings, Discussions

"Technology Horizons" also made use of a broad range of inputs gained from numerous site visits, briefings, and discussions involving organizations within the Air Force and elsewhere in the Department of Defense, as well as other federal agencies, federally-funded research and development centers (FFRDCs), national laboratories, and companies. A partial list of these organizations is given below.

Air Staff, Headquarters Air Force (HAF)

- Manpower and Personnel (AF/A1)
- Intelligence, Surveillance and Reconnaissance (AF/A2)
- Operations, Plans and Requirements (AF/A3/5)
- Logistics, Installations and Mission Support (AF/A4/7)
- Strategic Plans and Programs (AF/A8)
- Studies, Assessments and Lessons Learned (AF/A9)
- Strategic Deterrence and Nuclear Integration (AF/A10)
- Air Force Historian (AF/HO)
- Air Force Reserve (AF/RE)
- Air Force Safety (AF/SE)
- Air Force Chief Scientist (AF/ST)
- Air Force Surgeon General (AF/SG)

- Air Force Test and Evaluation (AF/TE)
- Air Force Scientific Advisory Board (AF/SB)
- CSAF Strategic Studies Group

Air Force Secretariat, Headquarters Air Force (HAF)

- Administrative Assistant (SAF/AA)
- Acquisition (SAF/AQ)
- Small Business Programs (SAF/SB)
- General Counsel (SAF/GC)
- Installations, Environment and Logistics (SAF/IE)
- Public Affairs (SAF/PA)
- Warfighting Integration and Chief Information Officer (SAF/XC)

Air Force Major Commands (MAJCOMs)

- Air Combat Command (ACC)
- Air Force Materiel Command (AFMC)
- Air Force Space Command (AFSPC)
- Air Education and Training Command (AETC)
- Air Force Reserve Command (AFRC)
- Air Force Special Operations Command (AFSOC)
- Air Mobility Command (AMC)
- Air National Guard (ANG)
- Pacific Air Forces (PACAF)

Air Force Product Centers

- Air Armament Center (AAC)
- Electronic Systems Center (ESC)
- Space and Missile Systems Center (SMC)

Direct Reporting Units (DRUs)

- U.S. Air Force Academy (USAFA)

Field Operating Agencies (FOAs)

- Air Force ISR Agency (AFISRA)
- Air Force Medical Support Agency (AFMSA)

Other Air Force

- Air Force Arnold Engineering Development Center (AEDC)
- Air Force Flight Test Center (AFFTC)
- Air Force Institute of Technology (AFIT)
- Air Force Research Laboratory (AFRL)
- Air Force Unmanned Systems ISR Innovations (AF/A2U)
- Air Force Warfare Center (USAFWC)
- Air Force Nuclear Weapons Center (AFNWC)
- Air Force A8 Futures Game 2009
- Distributed Mission Operations Center (DMOC)
- Space Innovation and Development Center (SIDC)
- National Air and Space Intelligence Center (NASIC)
- Oklahoma City Air Logistics Center (OC–ALC)
- 8th Air Force (8AF)
- 505th Command and Control Wing (505CCW)
- 688th Information Operations Wing (688IOW)

- 67th Network Warfare Wing (67NWW)
- 46th Test Wing (46TW)

Federal Agencies

- Defense Advanced Research Projects Agency (DARPA)
- Defense Threat Reduction Agency (DTRA)
- National Aeronautics & Space Administration (NASA)
- Office of Science & Technology Policy (OSTP)
- Federally Funded Research & Development Centers (FFRDCs)
- RAND Corporation
- Lincoln Laboratory

National Laboratories

- Argonne National Laboratory
- Lawrence Livermore National Laboratory
- Sandia National Laboratories

Companies

- Aernnova
- Aerojet
- Astrox
- Blue Origin
- Boeing Integrated Defense Systems
- General Atomics Aeronautical Company
- General Atomics Photonics Division
- Honeywell Aerospace
- Lockheed Martin Aeronautics Company
- Northrop Grumman Aerospace Systems
- Pratt & Whitney
- Raytheon Company
- Rolls Royce Liberty Works
- VMware

Other

- Air Force Red Team
- Air Force Studies Board (AFSB)
- Office of Net Assessment (ONA)
- Office of Naval Research (ONR)
- Naval Research Laboratory (NRL)

B.3 Studies, Reports, and Other Documents

[1] Quadrennial Defense Review Report. Office of the U.S. Secretary of Defense, 2010.

[2] Joint Operating Environment. Joint Forces Command, U.S. Secretary of Defense, 2010.

[3] Global Trends 2025: A Transformed World. National Intelligence Council, NIC 2008-003, 2008.

[4] Mapping the Global Future: Report of the National Intelligence Council's 2020 Project. National Intelligence Council, NIC 2004-13, 2004.

[5] Joint Warfighting Science and Technology Plan. Director of Defense Research and Engineering, U.S. Department of Defense, 2008.

[6] The National Security Strategy of the United States of America. Office of the President of the United States, 2006.

[7] Air Force Roadmap 2006-2025. U.S. Air Force, AFD-060713-002, 2006.

[8] Focused Long Term Challenges (FLTCs): Capability Visions / Problem Statements / Technology Challenges Definitions. Baseline Plan. HQ AFMC/A2/5, 2007.

[9] US Air Force Futures Capabilities Game 2007 Unclassified Report Summary. Directorate of Strategic Planning, Headquarters U.S. Air Force, 2008.

[10] Operational Impact of Exponential Technological Change on the Air Force. Blue Horizons 2007, Center for Strategy & Technology, Air University, 2008.

[11] Blue Horizons 2007, "Horizon 21" Project Report. Air University Center for Strategy and Technology, Air University, Air Education and Training Command (AETC), 2007.

[12] Air Force 2025: A Future Study Conducted 1995-1996 for the Air Force Chief of Staff. Air University, Air Education and Training Command (AETC), 1996.

[13] New World Vistas: Air and Space Power for the 21st Century, Vols. 1-13. USAF Scientific Advisory Board, 1995.

[14] Toward New Horizons: A Report to the General of the Army H.H. Arnold. Army Air Forces Scientific Advisory Group (Theorode von Kármán, ed.), 1945.

[15] Spacecast 2020: A Future Study Conducted 1994-1994 for the Air Force Chief of Staff. Air University, Air Education and Training Command (AETC), 1994.

[16] America's Strategic Posture: The Final Report of the Congressional Commission on the Strategic Posture of the U.S. Perry, W. et al., U.S. Institute of Peace Press, 2009.

[17] 21st Century Strategic Technology Vectors: Volume 1 – Main Report; Volume 2 – Critical Capabilities and Enabling Technologies; Volume 3 – Strategic Technology Planning; Volume 4 – Accelerating the Transition of Technologies into U.S. Capabilities. Defense Science Board (DSB), 2007.

[18] The Global Technology Revolution 2020 – In-Depth Analyses: Bio / Nano / Materials / Information Trends, Drivers, Barriers, and Social Implications. Silberglitt, R. et al, RAND National Security Research Division, 2006.

[19] Science and Engineering Indicators 2010. National Science Board, National Science Foundation, 2010.

[20] Science and Engineering Indicators 2008: Volume 1 (NSB 08-01), Volume 2 (NSB 08-01A); National Science Board, National Science Foundation, 2008.

[21] Capability Surprise. Defense Science Board (DSB), 2008 Summer Study, 2009.

[22] Air Force Research Laboratory Capability-Based Science & Technology Strategy 2030. Headquarters, Air Force Research Laboratory, U.S. Air Force, 2007.

[23] The Challenges to US National Security; Strategy for the Long Haul Series. Krepinevich, A., Martinage, R., and Work, R., Center for Strategic and Budgetary Assessments, 2008.
[24] The Future Security Environment: Multidimensional Challenges in a Multi-Player World. Work, R., Center for Strategic and Budgetary Assessments (CBSA), 2008.
[25] Challenges to Military Operations in Support of U.S. Interests: Volume 1 – Executive Summary, Volume 2 – Main Report. Defense Science Board (DSB), 2008.
[26] An Air Force Strategy for the Long Haul; Strategy for the Long Haul Series. Ehrhard, T., Center for Strategic and Budgetary Assessments (CBSA), 2009.
[27] Air Force Strategy to Resources. Deputy Chief of Staff, Strategic Plans and Programs, Headquarters, U.S. Air Force, 2008.
[28] Shaping the Future Air Force. Shlapak, D. et al, RAND Project Air Force, 2006.
[29] Avoiding Surprise in an Era of Global Technology Advances. Committee on Defense Intelligence Agency Technology Forecasts & Reviews, National Research Council, 2005.
[30] Dialogue 1: The Changing Global Landscape for Technology Leadership. Council on Competitiveness, 2009.
[31] Positioning, Navigation and Timing Science and Technology Roadmap. Office of the Secretary of Defense, 2005.
[32] USAF Intelligence, Surveillance, & Reconnaissance Flight Plan: The Key Guidance for Planning the Current & Future ISR Force. Headquarters, United States Air Force, 2009.
[33] Policy and Purpose: The Economy of Deterrence. Gen N.A. Schwartz and Lt T.R. Col Kirk, Strategic Studies Quarterly, Air University Press, 2009.
[34] DoD Science and Technology: Strategy for the Post-Cold War Era. Beason, J., Industrial College of the Air Force, National Defense University Press, 1997.
[35] Lead Turning the Future: 2008 Strategy for U.S. Air Force ISR. Deputy Chief of Staff for Intelligence, Surveillance and Reconnaissance, Headquarters, U.S. Air Force, 2008.
[36] Air Combat Command Enabling Concept for Directed Energy. Air Combat Command (ACC), U.S. Air Force, 2009.
[37] China's Air Defence Missile Systems. Kopp, C., Defence Today, 2008.
[38] Post Cold War Air-to-Air Missile Evolution. Kopp, C., Defence Today, 2009.
[39] Recent U.S. and Chinese Antisatellite Activities. LtCol J. Mackey, Air and Space Power Journal, Vol. 23 (9), 2009.
[40] The Art of Military Discovery: Chinese Air and Space Power Implications for the USAF. Strategic Studies Quarterly, 2010.
[41] Day Without Space. Air Combat Command (ACC), 2009.
[42] The Value of Stealth. Air Combat Command (ACC), 2009.
[43] Air Force Energy Plan 2010. Office of the Secretary of the Air Force, U.S. Air Force, 2010.
[44] Return of the Bomber: The Future of Long-Range Strike. Grant, R., Air Force Association (AFA), 2007.
[45] Back From the Future: The Impact of Change on Airpower in the Decades Ahead. Clodfelter, M., Strategic Studies Quarterly, Air University Press, 2009.
[46] The Nation's Guardians: America's 21st Century Air Force. CSAF White Paper, Gen T.M. Moseley, Chief of Staff, U.S. Air Force, 2007.

[47] Global Reach – Global Power: The Evolving Air Force Contribution to National Security. Aerospace Education Foundation, DTIC 19970610 059, 1992.

[48] Harnessing the Genie: Science and Technology Forecasting for the Air Force 1944-1986. Gorn, M., Office of Air Force History, 1988.

[49] Report of the Defense Science Board Task Force on Time Critical Conventional Strike for Strategic Standoff. Defense Science Board (DSB), 2009.

[50] The Anti-Access Threat and Theater Air Bases. Bowie, C., Center for Strategic and Budgetary Assessments (CBSA), 2002.

[51] Meeting the Anti-Access and Area-Denial Challenge. Krepinevich, A., Watts, B. and Work, R., Center for Strategic and Budgetary Assessments (CBSA), 2003.

[52] Alternatives for Long-Range Ground-Attack Systems. Congressional Budget Office (CBO), 2006.

[53] Cyberdeterrence and Cyberwar. Libicki, M. C., RAND Project Air Force, 2009.

[54] Game Changer Concepts. Marvin, D.C., Research and Program Development Office, The Aerospace Corporation, 2007.

[55] Air Defense Against Unmanned Aerial Vehicles: Vol. 1 – Executive Summary & Annotated Brief, Vol. 2 – Final Report. USAF Scientific Advisory Board, SAB-TR-06-01, 2006.

[56] Persistence at Near Space Altitudes. USAF Scientific Advisory Board, SAB-TR-05-06, 2005.

[57] Chip-Scale Atomic Devices: Precision Instruments based on Lasers, Atoms and MEMS. Kitching, J., National Institute of Standards and Technology (NIST), 2009.

[58] The Cruise Missile Challenge. Mahnken, T., Center for Strategic & Budgetary Assessments (CBSA), 2005.

[59] Implications of Cyber Warfare: Vol. 1– Executive Summary & Annotated Brief, Vol. 2– Main Report. USAF Scientific Advisory Board, SAB-TR-07-02, 2007.

[60] Implications of Spectrum Management for the Air Force. USAF Scientific Advisory Board, SAB-TR-08-03, 2008.

[61] Kinetic Precision Effects. USAF Scientific Advisory Board, SAB-TR-08-02, 2008.

[62] Operational Utility of Small Satellites. USAF Scientific Advisory Board, SAB-TR-07-03, 2007.

[63] Assuring Access in Key Strategic Regions: Towards a Long-Term Strategy. Larson, E. et al, RAND Arroyo Center, 2004.

[64] Technology Options for Improved Air Vehicle Fuel Efficiency. USAF Scientific Advisory Board, SAB-TR-06-04, 2006.

[65] Wired for War: The Robotics Revolution and Conflict in the 21st Century. Singer, P.W., Penguin Press, 2009.

[66] Prophecy Fulfilled: "Toward New Horizons" and Its Legacy. (Gore, M. ed.), Air Force History and Museums Program, U.S. Air Force, 1994.

[67] Defense Science Board Task Force on Directed Energy Weapons. Defense Science Board (DSB), 2007.

[68] The E-Bomb: How America's New Directed Energy Weapons will Change the Way Future Wars will be Fought. Beason, D., Da Capo Press, 2005.

[69] Chip-Scale Atomic Clock. Nguyen, C., Microsystems Technology Office (MTO), Defense Advanced Research Projects Agency (DARPA), 2005.

[70] NGAS Lasercom Technology. Aerospace Systems Sector, Northrup Grumman, 2009.

[71] Ultra-Cold Atoms and Precision Inertial Navigation Systems. Precision Inertial Navigation Systems (PINS), Defense Science Office (DSO), DARPA, 2010.
[72] Report of the Defense Science Board Task Force on Strategic Communication. Defense Science Board (DSB), 2008.
[73] Technologies of Choice for Future Space Investment. Ewart, R., Space and Missile Systems Center, Air Force Space Command (AFSPC), 2009.
[74] Converging Technologies for Improving Human Performance: Nanotechnology, Biotechnology, Information Technology and Cognitive Science (Roco, M. and Bainbridge, W. S. eds.), National Science Foundation, 2002.
[75] Capability of the People's Republic of China to Conduct Cyber Warfare and Computer Network Exploitation. Krekel, B., Information Systems, Northrop Grumman, 2009.
[76] Use and Sustainment of Composites in Aircraft. USAF Scientific Advisory Board, SABTR-07-04, 2007.
[77] A Scientific Research and Development Approach to Cyber Security. (Catlett, C. ed.), submitted to U.S. DoE on behalf of the Research and Development Community, 2008.
[78] Future Need for VTOL/STOL Aircraft. Defense Science Board (DSB), 2007.
[79] Spacelift Development Plan. SMC Development Planning, SMC Launch and Range Systems Wing, Air Force Space Command (AFSPC), 2009.
[80] Operationalizing Space Launch. USAF Scientific Advisory Board, SAB-TR-04-03, 2004.
[81] Australia and Cyber-Warfare. Water, G., Ball, D. and Dudgeon, I., The Australian National University, 2008.
[82] An Overview of Cloud Computing: The Future is Virtually Here. Andrews, A. E. and Segal, M. E., National Security Agency, The Next Wave, Vol. 17, No. 4, 2009.
[83] China, Space Weapons, and U.S. Strategic Posture. MacDonald, B., Strategic Posture Review Commission, 2009.
[84] Defence Technology Strategy for the Demands of the 21st Century. UK Ministry of Defence, 2006.
[85] Future Character of Conflict. Strategic Trends Program, UK Ministry of Defence, 2010.
[86] Defending Australia in the Asia Pacific Century: Force 2030. Defence White Paper. Australian Government Department of Defence, 2009.
[87] National Plan for Aeronautics Research and Development and Related Infrastructure. National Science and Technology Council, 2007.
[88] Autonomous Military Robotics: Risk, Ethics, and Design. Lin, P., Bekey, G. and Abney, K., Office of Naval Research (ONR), 2008.
[89] Options for Strategic Military Transportation Systems. Congressional Budget Office (CBO), 2005.
[90] Matching Resources with Requirements: Options for Modernizing the US Air Force. Kosiak, S., Center for Strategic & Budgetary Assessments (CBSA), 2004.
[91] Spending on US Strategic Nuclear Forces: Plans & Options for the 21st Century. Kosiak, S., Center for Strategic & Budgetary Assessments (CBSA), 2006.
[92] Future Warfare 20XX Wargame Series: Lessons Learned Report. Vickers, M. and Martinage, R., Center for Strategic & Budgetary Assessments (CBSA), 2001.
[93] The Military Use of Space: A Diagnostic Assessment. Watts, B., Center for Strategic & Budgetary Assessments (CBSA), 2001.

[94] Alternatives for Future U.S. Space-Launch Capabilities. Congressional Budget Office (CBO), 2006.
[95] Why Should DoD Invest in Basic Research? – A Presentation for GOMAC Tech 2008. Rees, W., Deputy Undersecretary of Defense (Laboratories and Basic Sciences), 2008.
[96] Net Assessment 2008 Summer Study. Portfolio Group, Director of Net Assessment in the Office of the Secretary of Defense, 2008.
[97] Options for Operating USAF Bases Under Attack. Stillion, J., Caston, L. and Vick, A., RAND Project Air Force, 2008.
[98] Air Power in the New Counterinsurgency Era: The Strategic Importance of USAF Advisory and Assistance Missions. Vick, A. et al, RAND Project Air Force, 2006.
[99] The Revolution in War. Vickers, M. and Martinage, R., Center for Strategic & Budgetary Assessments (CBSA), 2004.
[100] Forecasting the Progress of Technology, Maj Martino, J., Air University Review, 1969.
[101] Thermal Management Technology Solutions. USAF Scientific Advisory Board, SAB-TR07-05, 2007.
[102] Developing Science and Technologies List (DSTL), Military Critical Technologies Program, U.S. Department of Defense.
[103] Military Critical Technologies List (MCTL), Military Critical Technologies Program, U.S. Department of Defense.
[104] Future Air Force Needs for Survivability. Air Force Studies Board, National Research Council, National Academy of Science, 2006.
[105] Alternatives for Military Space Radar. Congressional Budget Office (CBO), 2007.
[106] Report of the Defense Science Board Task Force on Understanding Human Dynamics. Defense Science Board (DSB), 2009.
[107] Creating an Assured Joint DOD and Interagency Interoperable Net-Centric Enterprise. Defense Science Board (DSB), 2009.
[108] Operations Research Applications for Intelligence Surveillance & Reconnaissance (ISR). Defense Science Board, 2009.
[109] Wideband Radio Frequency Modulation: Dynamic Access to Mobile Information Networks. Defense Science Board (DSB), 2003.
[110] On n^{th} Order Attacks. Bilar, D., Department of Computer Science, University of New Orleans, 2009.
[111] Applying Electronic Warfare Solutions to Network Security. Maj R. Smith, R. and Knight, S., *Canadian Military Journal*, Autumn 2005.
[112] Memorandum for CSAF: Strategic Concerns the Deserve CSAF Attention. U.S. Air Force Futures Group, 2007.
[113] Chinese and American Network Warfare. Thomas, T. L., Joint Forces Quarterly, 38, 2009.
[114] Seeding the Clouds: Key Infrastructure Elements for Cloud Computing. IBM, 2009.
[115] Warfighting in Cyberspace. Alexander, K. B., Joint Forces Quarterly, 46, 2007.
[116] A Review of United States Air Force and Department of Defense Aerospace Propulsion Needs. Air Force Studies Board, National Research Council, 2006.
[117] The Long-Term Implications of Current Plans for Investment in Major Unclassified Military Space Programs. Congressional Budget Office (CBO), 2005.
[118] System-Level Experimentation. USAF Scientific Advisory Board, SAB-TR-06-02, 2006.

[119] Guideposts for the U.S. Military in the Twenty-First Century. (Neufeld, J. ed.), Air Force History and Museums Program, U.S. Air Force, 2000.

[120] The United States Air Force and the Culture of Innovation: 1945-1965. Johnson, S., Air Force History and Museums Program, U.S. Air Force, 2002.

[121] Architects of American Air Supremacy: General Hap Arnold and Dr. Theodore von Karman. Maj D. Daso, Air University Press, 1997.

[122] Cyberspace: The New Battleground. Hull, G., Northrup Grumman, 2009.

[123] Proprietary Report for ISR Application of RBCC. TR-09121, Astrox Corporation, 2009.

[124] Weapons File 2009. Air Armament Center (AAC), Air Force Materiel Command (AFMC), 2009.

[125] National Aeronautics RDT&E Infrastructure Plan: Final IIWG Draft. Aeronautics RDT&E Infrastructure Interagency Working Group, 2009.

[126] Air Force Research Laboratory Business Plan FY08. Headquarters, Air Force Research Laboratory (AFRL), 2007.

[127] The Unmanned Imperative. Chief of Naval Operations Strategic Studies Group XXVIII, 2009.

[128] The Next 100 Years: A Forecast for the 21st Century. Friedman, G., Doubleday, 2008.

[129] Air Force Research Laboratory Strategic Management Plan. Headquarters, Air Force Research Laboratory (AFRL), 2007.

[130] The Future of War: Power, Technology and American World Dominance in the Twenty-First Century. Friedman, G., St. Martin's Press, Griffen Publ., 1998.

[131] Modern Strategy. Gray, C., Oxford University Press, 1999.

[132] Making Twenty-First-Century Strategy: An Introduction to Modern National Security Processes and Problems. Drew, D. and Snow, D., Air University Press, 2006.

[133] Why Air Forces Fail: The Anatomy of Defeat. (Higham, S. and Harris, S. eds), University of Kentucky Press, 2006.

[134] Addressing the Trust Asymmetry Problem in Grid Computing with Encrypted Computation. Dinda, P. A., Northwestern University, 2009.

[135] Advanced PRSEUS Structural Concept Design and Optimization. Li, V. and Velicki, A., American Institute of Aeronautics & Astronautics, AIAA 2008-5840, 2008.

[136] Army Energy Security Implementation Strategy. Army Senior Energy Council and Office of the Deputy Assistant Secretary of the Army for Energy and Partnerships, 2009.

[137] Military Compensation: Requirements, Trends and Options. Kosiak, S., Center for Strategic & Budgetary Assessments (CBSA), 2005.

[138] Conventional Arms Transfers to Developing Nations: 2001-2008. Grimmett, R., Congressional Research Service, 2009.

[139] Software Application Security. Stytz, M. and Hughes, J., Software Protection Initiative, U.S. Department of Defense, 2009.

[140] A Cooperative Strategy for 21st Century Seapower. Commandant of the Marine Corps, Chief of Naval Operations, and Commandant of the Coast Guard, 2007.

[141] Seven Deadly Scenarios: A Military Futurist Explores War in the 21st Century. Krepinevich, A.F., Bantum Books, 2009.

[142] Future Warfare Anthology. Maj Gen R. Scales, Jr. ed, U.S. War College, 2000.

[143] Air and Space Power in the New Millennium. (Goure, D. and Szara, C. eds), Center for Strategic & International Studies, 1997.

[144] Air Combat Past, Present and Future. Stillion, J. and Perdue, S., RAND Project Air Force, 2008.

[145] On "Other War": Lessons from Five Decades of RAND Counterinsurgency Research. Long, A., RAND National Defense Research Institute, 2006.

[146] Aerospace Operations Against Elusive Ground Targets. Vick, A.J. et al, MR-1398-AF, RAND Project Air Force, 200x.

[147] Evaluation of Options for Overseas Combat Support Basing. Amouzegar, M. et al, RAND Project Air Force, 2006.

[148] Air Force Phase Zero Capabilities-Based Analyses: Functional Area Analysis and Functional Needs Analysis. Stillion, J. et al, RAND Project Air Force, 2007.

[149] Military Operations Against Terrorist Groups Abroad: Implications for the United States Air Force. Ochmanek, D., RAND Project Air Force, 2003.

[150] Courses of Action for Enhancing U.S. Air Force Irregular Warfare Capabilities: A Functional Solution Analysis. Mesic, R. et al., RAND Project Air Force, 2008.

[151] Courses of Action for Enhancing USAF's Contributions to "Irregular Warfare" Operations. RAND/STR Team, RAND Project Air Force, 2008.

[152] Dissuading Terror: Strategic Influence and the Struggle Against Terrorism. Cragin, K. and Gerwehr, S., RAND Corporation, 2005.

[153] Security and Complexity: Are We on the Right Road. Hull, G., Northrup Grumman, 2009.

[154] The Future of Everything: The Science of Prediction. Orrell, D., Thunder's Mouth Press, 2007.

[155] The Next Fifty Years: Science in the First Half of the Twenty-First Century. (Brockman, J., ed), Vintage Press, 2002.

[156] The World is Flat 3.0: A Brief History of the Twenty-First Century. Freidman, T. L., Picador, 2007.

[157] Fourth-Generation War and Other Myths. Eschevarria, A., Strategic Studies Institute, U.S. Army, 2005.

[158] Insurgency and Counterinsurgency in the 21st Century: Reconceptualizing Threat and Response. Metz, S. and Millen, R., Strategic Studies Institute, U.S. Army, 2004.

[159] Future Warfare Anthology, Revised Edition. Maj Gen R. Scales, Jr., Strategic Studies Institute, U.S. Army, 2000.

[160] Rethinking Asymmetric Threats. Blank, S., Strategic Studies Institute, U.S. Army, 2003.

[161] Strategic Challenges for Counterinsurgency and the Global War on Terrorism. Murray, E. editor, Strategic Studies Institute, U.S. Army, 2006.

[162] Strategic Planning and the Drug Threat. Mendel, W. and Munger, M., Strategic Studies Institute, U.S. Army, 1997.

[163] Human Performance. JASON Report JSR-07-625, 2008

[164] Converging Technologies for Improving Human Performance: Nanotechnology, Biotechnology, Information Technology and Cognitive Science. (M.C. Roco et al, Eds.), National Science Foundation & Department of Commerce Report, 2002.

[165] Human Performance Enhancement. Brown, L. and Tvaryanas, A., *Air & Space Power Journal,* 2008.

[166] Proceedings of the 1st Augmented Human International Conference. ISBN 978-1-60558825-4, 2010.
[167] Transformation and Strategic Surprise. Gray, C., Strategic Studies Institute, U.S. Army, 2005.
[168] Transnational Threats: Blending Law Enforcement and Military Strategies. (Pumphrey, C. ed), Strategic Studies Institute, U.S. Army, 2000.
[169] The Accidental Guerrilla: Fighting Small Wars in the Midst of a Big One. Kilcullen, D., Oxford University Press, 2009.
[170] Lightning Rod: A History of the Air Force Chief Scientist's Office. Day, D.A., U.S. Air Force, 2000.
[171] Golden Legacy, Boundless Future: Essays on the United States Air Force and the Rise of Aerospace Power. (Cameron, R., Wittig, B. eds), Air Force History and Museums, 2000.
[172] From Reform to Reduction: Management of Navy and Department of Defense Labs in the Post-Cold War Era. Hazell, E., National Defense University & Naval Historical Ctr., 2008.
[173] Advanced Structural Concept Development Using Stitched Composites. Velicki, A. and Thrash, P., American Institute of Aeronautics & Astronautics, AIAA 2008-2329, 2008.
[174] Flying Blind: The Rise, Fall, and Possible Resurrection of Science Policy Advice in the United States. Kelly, H. et al., Federation of American Scientists, 2004.
[175] Technologies and Tactics for Improved Air-Ground Effectiveness. Jacobs, J. et al, RAND Project Air Force, MG-573-AF, 2008.
[176] Thinking about America's Defense: An Analytical Memoir. Kent, G.A. et al, RAND Project Air Force, 2008.
[177] Lest We Remember: Cold Boot Attacks on Encryption Keys. Halderman, J.A. et al, Proceedings 2008 USENIX Security Symposium, 2008.
[178] Security Guide for Critical Areas of Focus in Cloud Computing. Cloud Security Alliance, 2009.
[179] Taxonomy for the Modeling and Simulation of Emergent Behavior Systems. Holland, O. T., Naval Surface Warfare Center, SpringSim, Vol. 1, 2009.
[180] Infosphere: Smart Delivery of Fresh Information. Pu, C., Georgia Tech, 2000.
[181] Cybersecurity Strategies: The QuERIES Methodology. Carin, L, Cybenko, G. and Hughes, J., IEEE Computer Society, Computer, 2008.
[182] Petascale Computational Systems: Balanced Cyber Infrastructure in a Data-Centric World. Bell, G., Gray, J. and Szalay, A., Microsoft Research and Johns Hopkins University, 2005.
[183] Effectiveness of Air Force Science & Technology Program Changes. National Research Council (NRC), National Academy of Sciences, 2003.

B.4. Additional Inputs

In addition to the sources of inputs listed above, "Technology Horizons" sought further inputs from numerous sources to obtain the broadest range of opinions and viewpoints in formulating the findings and recommendations contained herein, and additionally made use of various classified sources as listed in Volume 3.

INDEX

#

A

B

D

E

T

U

V

W

Y